TRANSACTIONS

of the

American Philosophical Society

Held at Philadelphia for Promoting Useful Knowledge

VOLUME 74, Part 4, 1984

In Defense of the Earth's Centrality and Immobility:

Scholastic Reaction to Copernicanism in the Seventeenth Century

EDWARD GRANT

Professor of History and Philosophy of Science,
Indiana University—Bloomington

THE AMERICAN PHILOSOPHICAL SOCIETY

Independence Square, Philadelphia

1984

Library of Congress Catalog
Card Number 84–71078
International Standard Book Number 0–87169–744–0
US ISSN 0065–9746

CONTENTS

INTRODUCTION

In shifting from a geocentric to a heliocentric cosmology, Nicholas Copernicus earned a place among the truly great figures in the history
of science. His justly deserved reputation as a seminal thinker in
human history rests primarily on his attribution to the earth of a real daily
axial motion and an annual orbital motion around the sun—the latter
motion making a planet of the earth. Not only did Copernicus proclaim
the physical reality of these motions, but, of equal importance, furnished
the technical astronomical details that led ultimately to the abandonment
of the traditional Aristotelian-Ptolemaic cosmology and astronomy, which
had formed the basis of the medieval world view.

But even if Copernicus had written and said nothing about astronomical
and cosmological matters, his contemporaries and successors in the sixteenth
and seventeenth centuries would nonetheless have considered the possibility of the earth's axial and orbital motions.[1] During the late fifteenth
and sixteenth centuries, new sources became available to supplement those
that had been known in the Middle Ages, which learned of both possible
motions from Aristotle's *De caelo*,[2] and of the earth's possible axial rotation
from Ptolemy's *Almagest*,[3] Seneca's *Natural Questions*,[4] and Simplicius's
Commentary on Aristotle's De caelo.[5] Most of these arguments on the earth's

---

[1] Except for the tradition associated with Aristarchus of Samos, the orbital motions of the
earth described in the ancient world and usually linked with the Pythagoreans were not
heliocentric but were either around a geometric point or a central fire, the latter quite distinct
from the sun.

[2] In *De caelo* 2.13.293b.16–32 and 2.14.296a.24–26. In the latter passage Aristotle declared
that some make the earth "one of the stars, whereas others put it at the centre but describe
it as winding and moving about the pole as axis" (translation by W. K. C. Guthrie in the
Loeb Classical Library [London: William Heinemann Ltd; Cambridge, Mass.: Harvard University Press, 1960].

[3] See bk. 1, ch. 7 in R. Catesby Taliaferro's translation in *Great Books of the Western World*
(Chicago: Encyclopaedia Britannica, 1952) 16: 10–12.

[4] See vii. 2.3. The passage is translated by Thomas Heath, *Aristarchus of Samos, The Ancient
Copernicus* (Oxford: Clarendon Press, 1913), 308.

[5] Simplicius even named Heraclides of Pontus and Aristarchus of Samos as actual proponents
of the earth's axial rotation. Simplicius's important Greek commentary was translated into
Latin by William of Moerbeke in 1271 and printed in 1540 under the title *Simplicii philosophi
acutissimi commentaria in quatuor libros De celo Aristotelis Guillermo Morbeto interprete* (Venice,
1540). Almost immediately after its translation, Thomas Aquinas used it in his commentary
on *De caelo*, as is evident by his mention of both Heraclides and Aristarchus (for the passage,
see Edward Grant, ed., *A Source Book in Medieval Science* [Cambridge, Mass.: Harvard University
Press, 1974], 499–500, and n. 22). Nicole Oresme, who produced the most brilliant medieval
discussion of the earth's possible rotation, also mentioned Heraclides of Pontus (but not
Aristarchus). See *Nicole Oresme Le Livre du ciel et du monde*, ed. Albert D. Menut and Alexander
J. Denomy, translated with an Introduction by Albert D. Menut (Madison, Wis.: University
of Wisconsin Press, 1968), 521 and Grant, *Source Book*, 504, where Menut's translation is
reprinted.

motions, and others that had been proposed during the fourteenth century, were incorporated into Albert of Saxony's *Questions on De caelo,* which appeared in at least six editions between 1481 and 1520 and was widely cited by scholastic authors during the sixteenth and seventeenth centuries.[6]

Of the sources that became available in the sixteenth century, Copernicus himself (in the preface to the *De revolutionibus orbium coelestium*) cited two; namely, Cicero's *Academica* (bk. 2, 39, 123), where the Syracusan Hicetas is said to have assumed the earth's axial rotation and the immobility of the planets and stars,[7] and Plutarch's *De placitis philosophorum* (bk. 3, ch. 13), where Philolaus the Pythagorean is said to have ascribed an orbital motion to the earth and Heraclides of Pontus and Ecphantus the Pythagorean assigned to it a west to east axial rotation.[8] In his treatise *Concerning the Face Which Appears in the Orb of the Moon (De facie quae in orbe lunae apparet),* Plutarch also found occasion to mention the earth's axial and orbital motions when he reported a charge of impiety brought by Cleanthes against Aristarchus of Samos, who was "disturbing the hearth of the universe because he sought to save ⟨the⟩ phenomena by assuming that the heaven is at rest while the earth is revolving along the ecliptic and at the same time . . . rotating about its own axis."[9]

Supporters of the traditional Aristotelian cosmology were as knowledgeable about these new references to possible terrestrial motion as were their opponents. The new sources were frequently mentioned in Aristotelian commentaries of the sixteenth and seventeenth centuries, especially in

---

[6] For a list of editions and manuscripts of Albert's treatise, see Charles H. Lohr, "Medieval Latin Aristotle Commentaries: Authors A-F," *Traditio* 23 (1967): 350. The edition that will be cited in this study is that of George Lokert with the title *Questiones et decisiones physicales insignium virorum: Alberti de Saxonia in octo libros Physicorum; tres libros De celo et mundo; duos lib. De generatione et corruptione; Thimonis in quatuor libros Meteororum; tres lib. De anima; Buridani in lib. De sensu et sensato; . . . Aristotelis. Recognitae rursus et emendatae summa accuratione et iudicio Magistri Georgii Lokert Scotia quo sunt Tractatus proportionum additi* (Paris, 1518). Relevant discussions on the earth's motions appear in bk. 2, questions 13, 23, and 24.

[7] For Cicero's "Hicetas," Copernicus incorrectly substituted "Nicetus." For the explanation, and for a translation of Cicero's statement, see Edward Rosen's discussion in *Nicholas Copernicus On the Revolutions,* ed. Jerzy Dobrycki; translation and commentary by Edward Rosen (Baltimore: The Johns Hopkins University Press, 1978), 341.

[8] Long after the time of Copernicus, the *De placitis philosophorum* was falsely attributed to Plutarch and regularly included in the latter's *Moralia* (in the traditional order, the five books of the *De placitis* were placed in the eleventh of the fifteen volumes comprising the *Moralia*). It is now ascribed to Aëtius, who probably lived in the second century A.D. but is otherwise unknown. Copernicus, who quoted the Greek text in his preface, probably used the Aldine edition printed in Venice in 1509 (Rosen, *Nicholas Copernicus On the Revolutions,* 342). In 1510, Johannes Laurentius published a Latin translation in Rome and by 1603, at the latest, Philemon Holland published an English translation (*The Philosophie commonlie called the Morals written by the learned Philosopher Plutarch of Chaeronea,* translated out of Greeke into English . . . by Philemon Holland of Coventrie, Doctor in Physicke . . . at London by Arnold Hatfield, 1603).

[9] The translation is by Harold Cherniss and appears in Plutarch's *Moralia* with an English translation by Harold Cherniss and William C. Helmbold, vol. 12, 920A–999B (Loeb Classical Library; London: William Heinemann Ltd; Cambridge, Mass.: Harvard University Press, 1957), 55.

commentaries on *De caelo*. The scope of the traditional medieval commentaries was thus considerably increased. Even had they never heard
of Copernicus, Aristotelians would have been obligated to take cognizance
of these claims.

But all available evidence indicates that it was Copernicus's arguments,
rather than the brief, unsupported fragmentary statements from the sources
cited above that eventually posed the real challenge to Aristotelian cosmology. With a few exceptions, of whom Christopher Clavius was one,
the arguments in favor of the earth's motions in the first book of the *De
revolutionibus* were slow to arouse Aristotelians or the Catholic Church.

Only at a much later date, when it became evident that this work of Copernicus
was not intended for mathematicians alone; when it became clear that the blow
to the geocentric and anthropocentric Universe was deadly; when certain of its
metaphysical and religious implications were developed in the writings of Giordano
Bruno, only then did the old world react

and attempt to suppress the new ideas of the universe by "the condemnation
of Copernicus in 1616 and of Galileo in 1632."[10]

In describing and assessing the struggle between the Copernican and
Aristotelian world views, modern scholars have focused their attention
on the Copernican system treating the Aristotelian arguments as repre-
sentative of the obstinate, reactionary opposition of biased theologians.
Aristotelian arguments are included for discussion only if perchance they
had been specifically refuted by Copernicans. No serious attempt has yet
been made to study for its own sake the Aristotelian system in its final
form before it succumbed to Newtonian cosmology and physics at the
end of the seventeenth century. The significance of such a study should
be evident from the fact that the Aristotelian system continued to hold
the allegiance of the overwhelming majority of the educated classes in
the seventeenth century, its final century as a credible system. My objective
here is to begin this essential study by investigating certain fundamental
tenets of Aristotelian cosmology, namely the centrality and immobility of
the earth. Within the class of Aristotelian defenders of the earth's centrality
and immobility scholastic Aristotelians will be of primary interest.

Before proceeding, something must be said about the frequently used
terms "Aristotelian" ("Aristotelianism") and "scholastic" ("scholasticism").
To take the latter first, scholasticism was a method for analyzing and
explicating texts either by systematic and sequential commentary or by
systematic formulation of questions based on a specific text, as, for example,
Questions on Aristotle's De caelo or *Questions on Aristotle's Physics*. Indeed
by the seventeenth century, the various subdivisions of Aristotle's natural

---

[10] Alexandre Koyré, *The Astronomical Revolution: Copernicus-Kepler-Borelli*, translated by
Dr. R. E. W. Maddison, F.S.A. (Paris: Hermann; London: Methuen; Ithaca, N.Y.: Cornell
University Press, 1973; original French edition, 1961), 17.

philosophy were sometimes presented in a more integrated manner in treatises titled *cursus philosophicus.* In one form or another, these scholastic methods were in use in institutions of higher learning from approximately 1200 to 1700. Although scholasticism was most intimately associated with the works of Aristotle, as a methodology it was also applied to other authors and texts, both theological and secular. Scholastic authors who were concerned with Aristotle's treatises in natural philosophy were either secular masters or theologians. Most of those with whom we shall be concerned in the seventeenth century belonged to the latter group.

But if scholasticism embraced much more than the study of Aristotle's works on natural philosophy, it is equally true that Aristotelianism extended far beyond the interests of scholastic commentators and interpreters. Aristotelianism, or the study of Aristotle's works as a guiding philosophy for understanding the physical and spiritual worlds, won partial or total allegiance and respect not only from scholastic thinkers, but also from many sixteenth and seventeenth–century humanists, emerging scientists, and from both Catholics and Protestants. In this study, however, we shall focus on Aristotelians who were primarily scholastic theologians interested in the problems of the earth's centrality and immobility. It is no small irony that whereas in the thirteenth and early fourteenth centuries, many theologians had opposed the introduction of Aristotle's natural philosophy as potentially dangerous to the Christian faith, by the seventeenth century, most had become firm defenders of Aristotle's philosophy and world view as these had developed over the centuries.

Despite what has already been said, no serious attempt will be made here to define terms like Aristotelian or Aristotelianism. By the seventeenth century Aristotelianism had absorbed so much from other intellectual currents and Aristotelians had become such a diverse group that no meaningful definition could be formulated that would embrace them all.[11] Although this will be apparent in the course of this study, it is also probable that whatever definition may someday be deemed adequate, it would include virtually all who play a role in these pages.

---

[11] For the diversity of Aristotelianism as a whole and the difficulties in categorizing Aristotelians, see Charles Schmitt, *Aristotle and the Renaissance* (Cambridge, Mass./London: Published for Oberlin College by Harvard University Press, 1983), ch. 1 ("Renaissance Aristotelianisms"), 10–33. According to Schmitt (p. 10), it is essential to speak of "Renaissance Aristotelianisms" because "the single rubric Aristotelianism is not adequate to describe the range of diverse assumptions, attitudes, approaches to knowledge, reliance on authority, utilization of sources, and methods of analysis to be found among the Renaissance followers of Aristotle." See also his earlier discussions in *A Critical Survey and Bibliography of Studies on Renaissance Aristotelianism, 1958–1969* (Saggi e testi, xi; Padua, 1971), 16–18 and "Towards a Reassessment of Renaissance Aristotelianism," *History of Science,* 11, pt. 3, nr. 13 (1973), 159–173. Before leaving the problem of terminology, the reader should be aware that authors labeled as "Scotists," "Jesuits," and "Carmelites," etc., are so designated solely for purposes of identification and not for any significant doctrinal differences. With respect to the earth's centrality and immobility, such labels are irrelevant.

I. THE DIVERSITY OF THE ARISTOTELIAN REACTION

The Aristotelian reaction and response to the heliocentric cosmology of the sixteenth and seventeenth centuries has too often been characterized as uniformly unimaginative and hostile. Among the defenders of the traditional cosmology, Catholic scholastic theologians are usually thought to have been the most hostile of all to the idea of assigning any motion whatever to the earth. Such a sweeping claim must, however, be viewed with suspicion, not only because there was at least one notable exception, as the description of Thomas White's opinions (below) will reveal, but also because a number of Aristotelians, both scholastic and nonscholastic, became convinced that contemporary astronomy demanded certain compromises concerning the traditional arrangement and order of the planets, as well as the behavior of the earth itself.

For an Aristotelian who possessed such convictions, adoption of some form of Tycho Brahe's geoheliocentric system was the most plausible course of action available. One of those who took this direction was Johann Heinrich Alsted (1588–1638), an influential Aristotelian natural philosopher who was a professor of philosophy and theology at the Protestant university of Herborn.[12] Alsted compared the basic hypotheses that supported the planetary theories of Copernicus, Brahe, Ptolemy "and the ancient philosophers," Nicholas Raymarus, and Helisaeus Röslin.[13] Among these numerous hypotheses, Alsted declared that "many are contrary to Holy Scripture, [and] many are contrary to experience, reason, and physical

---

[12] See Sister Mary Reif, "Natural Philosophy in Some Early Seventeenth Century Scholastic Textbooks" (Ph.D dissertation, St. Louis University, 1962), 9–10. In Charles Webster's judgment, Alsted favored "a 'Christianized' Peripatetic philosophy" ("Alsted, Johann Heinrich," *Dictionary of Scientific Biography* [16 vols.; New York: Charles Scribner's Sons, 1970–80), 1: 126.

[13] See Johann Heinrich Alsted, *Methodus admirandorum mathematicorum complectens novem libros matheseos universae* (Herborn, 1613). I am grateful to my student, Peter Lang, for calling Alsted's discussion to my attention. "Raymarus" is Nicolaus Reymers Baer (1550–1599), or Ursus. In 1588, Ursus published the *Fundamentum Astronomicum* in which he proposed a geoheliocentric system similar to that of Tycho Brahe, differing from the latter, however, in attributing a daily rotation to the earth and an orbit for Mars that enclosed, but did not intersect with the sun's orbit. Despite this difference, Tycho accused Ursus of plagiarism. For the controversy between them, see Christine Jones, "The Geoheliocentric Planetary System: Its Development and Influence in the Late Sixteenth and Seventeenth Centuries" (Ph.D. dissertation; University of Cambridge, 1964), 108–135 and Victor E. Thoren, "The Comet of 1577 and Tycho Brahe's System of the World," *Archives internationales d'histoire des sciences*, 29, no. 104 (1979): 62–66. "Röslin" is Helisaeus Roeslin, who first published his version of a geoheliocentric planetary system in 1597 (for details, see Jones, ibid., 136–144).

principles."[14] Röslin's have been proven and are therefore acceptable. Six in all,[15] they included the most fundamental of Aristotelian and Tychonic assumptions: the earth lies immobile at the center of the universe.[16] The other planetary dispositions were, however, Tychonic: the sun and moon circle the earth while the other five planets orbit the sun. But where Tycho denied the existence of planetary orbs, Röslin postulated them as the bearers of the planets[17] and thereby convinced Alsted of the superiority of his hypotheses.

Drastic as were the changes in traditional cosmology accepted by those Aristotelians who sought refuge in Tychonic astronomy, they, at least, did not abandon the seemingly inviolate principles of the earth's centrality and immobility. There were, however, Aristotelians who dared set the earth into one or more circular motions. In 1571, Andreas Cesalpino (1519–1603), who makes no mention of Copernicus, argued[18] that although the earth itself had only one natural motion, which, as Aristotle had argued, was rectilinear and directed toward the middle of the world, it could move with a circular motion if that motion was imposed on it by an external force. But whence could such a force derive? From the surrounding air and ultimately from the heavens. Cesalpino insisted[19] that as the heaven, which he envisioned as a single continuum, moved from east to west, its force also caused the elemental spheres of fire and air to move in the same direction, but at a much slower rate. The rotating sphere of air, in turn, would incessantly impact on the uneven and continually changing earth's surface and push or carry it in the same east to west direction. Cesalpino was impressed with the motive force of air, which could move massive ships merely by pressing on their sails. If small parts of air could move large ships, the whole mass of air moving from east to west ought to be

---

[14] Ibid., 38. Although he admired Copernicus (Webster, "Alsted," *DSB*, 1: 126), Alsted denied the earth's daily and annual motions and specifically repudiated the Copernican hypotheses (*Methodus admirandorum*, 33) as contrary to physical principles because they implied an enormous void space between the sphere of Saturn and the eighth sphere of the fixed stars. Presumably this gap was entailed by a lack of stellar parallax that required Copernicus to locate the starry orb much farther from the earth than was required in the Aristotelian-Ptolemaic system.

[15] *Methodus admirandorum*, 37.

[16] Some geoheliocentrists, such as Ursus, assumed the daily rotation of the earth. Indeed Longomontanus (i.e. Christian Severin [1562–1647]), Tycho's major disciple, also assumed it and thereby subscribed to what has come to be known as "the semi-Tychonic system" (see Thoren, "The Comet of 1577," *Archives internationales* 29 [1979]: 63, n. 47). It should be noted that, unlike Alsted, neither Ursus nor Longomontanus was a scholastic theologian and very likely neither was an Aristotelian natural philosopher. As we shall see below, there were Aristotelians who would allow the earth's rotation, but they were not supporters of Tycho.

[17] On Röslin's (Roeslin) insistence on solid, planetary orbs, see also Jones, *The Geoheliocentric Planetary System,* 138–141.

[18] *Andreae Cesalpini Aretini medici clarissimi atque philosophi subtilissimi peritissimique peripateticarum quaestionum libri quinque* (Venice, apud Iuntas, 1571), Bk. 3, question 4 ("Planetas in circulis non in sphaeris moveri"), fols. 53r–59v.

[19] Ibid., fols. 58v–59v.

capable of pushing the entire earth with a circular motion in the same direction,[20] an east to west motion that would be the slowest in the universe, since the earth was farthest removed from the eighth sphere of the fixed stars. But if the force of air was sufficient to cause the earth to move with the slowest circular motion in the universe, it lacked the power to cause that same earth to move rectilinearly from the center of the universe.[21] The earth's circular motion occurred while it was forever stationary at the center of the world. Thus did Cesalpino reconcile the earth's circular motion with the basic Aristotelian requirement that it lie immobile at the center of the world. By this means, moreover, Cesalpino sought to account for the precession and trepidation of the equinoxes, conceiving these celestial motions as mere appearances derived from the earth's circular motion. The earth's slow east to west motion gives to the sphere of fixed stars the appearance of a slow west to east motion, which would correspond to the motion of precession falsely ascribed to the sphere of the fixed stars. Because of its uneven, and continually changing, surface, however, the earth's east to west rotatory motion is irregular and unpredictable, thus producing an apparent trepidation in the stars.[22] In this extraordinary

---

[20] Indeed if the earth were not moved with, and by, the air, "the peaks of the highest mountains would be worn away by the continuous rotation of the air," a consequence that is not observed. ("Signum praeterea est moveri terram cum aere: nam si aeris cursum non consequeretur, altissimorum montium cacumina continua aeris rotatione attererentur." Ibid., fol. 59r.)

[21] "Ex centro enim dimoveri impossibile est ne minimum quidem, non enim aer huiusmodi impulsum praebet." Ibid., fol. 58v.

[22] Cesalpino argued for the attribution of only a single motion to the fixed stars, namely, the daily motion from east to west. The motions of precession and trepidation only appear to be a west to east motion in the stars because of the earth's east to west motion. Both precession and trepidation are a direct consequence of that single, slow movement of the earth from east to west. Cesalpino's sense of trepidation seems to differ from the traditional astronomical conception associated with Thabit ibn Qurra and the numerous astronomers who accepted it in the Latin West. The latter thought of it as a slow movement back and forth, that is, from west to east and east to west, and attempted to reconcile that oscillating motion with precession. For Cesalpino, however, "trepidation" is really equated with the irregularity of the earth's slow east to west velocity rather than a movement now in one direction and then in the opposite direction. For a brief description of precession and trepidation, see Olaf Pedersen and Mogens Pihl, *Early Physics and Astronomy* (London: MacDonald and Janes; New York: American Elsevier Inc., 1974), 183–185.

For this unusual opinion, Cesalpino was severely criticized by Raphael Aversa, who insisted that these motions were only appropriate to the heavens and not the earth. In Aversa's judgment, Cesalpino "shamefully erred" ("Cesalpinus in hoc turpiter lapsus fuit") when he proclaimed this opinion. See Aversa, *Philosophia metaphysicam physicamque complectens quaestionibus contexta* (2 vols.; Rome: apud Iacobum Mascardum, 1625, 1627), 2, question 34, sect. 5: 141, cols. 1–2 (for Aversa's description of Cesalpino's ideas) and 143, col. 1 (for his criticism). Hereafter this treatise will be cited as Aversa, *Philosophia*. In their *Disputations on De caelo,* Bartholomew Mastrius and Bonaventura Bellutus also denied Cesalpino's claim. Not only did Sacred Scripture assign all circular motions to the heavens and rest to the earth, but the earth has neither an internal capacity to move circularly nor is there any external force that could cause its continual rotation. None of the other elements, including air, has the power to move it. See *RR. PP. Bartholomaei Mastrii de Meldula et Bonaventurae Belluti . . . philosophiae ad mentem Scoti cursus integer . . . Editio novissima a mendis* (5 vols.; Venice: apud Nicolaum Pezzana, 1727; earlier editions in 1678, 1688, and 1708). The *Disputations*

manner, did Andreas Cesalpino assign to the earth a circular motion, explain certain astronomical phenomena by that motion, and yet remain faithful to the traditional Aristotelian conception of an earth located in the center of the universe but unable to move itself circularly by naturally inherent properties.

Not many Aristotelians would go beyond Cesalpino. But if Thomas White (1593–1676), an English Catholic and sometime professor of philosophy and theology, is included among them—and he seems to have proclaimed membership in the company of Aristotelians—then it is possible to think the unthinkable: the most basic elements of the Copernican theory, namely the earth's daily rotation and its annual motion around a stationary sun, are somehow reconcilable with Aristotelianism. The reconciliation appears in White's *Peripateticall Institutions*, where, in a section called "The Authour's Design," White offers the following explanation of the work's title: "Why I have stiled them Institutions, the shortnesse and concise connection of the work sufficiently discover. I call them Peripateticall because, throughout they [i.e. the 'Institutions,' or foundations] subsist upon Aristotle's principles; though the conclusions sometimes dissent."[23] In the concluding sentence, British understatement may have had its finest hour, since White's "dissent" led him to the assumption of the truth of the Copernican theory. In a manner reminiscent of Cesalpino, White, in Lessons XIV and XV, explained the earth's daily axial rotation by an east to west sweep of the wind which causes the upper part of the seas to begin a process that enables the lowest level of the sea in direct contact with the sea bed, or earth, to produce a west to east motion of the earth. The earth's daily circular motion is possible in this manner because it is not contrary to the earth's natural gravity and therefore offers no resistance to the west to east force of the seas at the points of contact.[24]

But, as White explained further, "because 'tis almost impossible this impulse should be equall on all sides, and cause a pure rotation about the Centre; there will, of necessity, a Progressive motion be mixt with it." But

---

on *De coelo* appears in the third volume of the 1727 edition and bears the title: *Tomus tertius: continens disputationes ad mentem Scoti in Aristotelis Stagiritae libros: De anima, De generatione et corruptione, De coelo, et Metheoris.* The discussions mentioned above appear on 563, col. 1. Subsequent references to this volume will take the form: Mastrius and Bellutus, *De coelo.* The various treatises included within this five–volume edition were first published as separate works between 1628 and 1640.

[23] *Peripateticall Institutions in the way of that eminent Person and excellent Philosopher Kenelm Digby. The Theoreticall Part. Also a Theological Appendix of the Beginning of the World.* By Thomas White, Gent. London, Printed by R. D. and are to be sold by John Williams at the sign of the Crown in S. Paul's Churchyard, 1656, sig. a4v–a5r. The work was originally published in Latin at Lyons in 1646. According to Phillip Drennon Thomas ("White, Thomas," *Dictionary of Scientific Biography,* 14: 301–302), White, an English Catholic, was "a devoted follower of Aristotle," although his "scientific treatises contain modifications and revisions of Aristotle's thought."

[24] *Peripateticall Institutions,* Lesson XIV, 174–175, paragraphs 1–3.

this motion, which represents the earth's annual orbit, must be "in one line" because "all the motions which Astronomers assign the Earth must, of necessity, compose one line; and, if the lashing or impulse of the underwater advance the Earth in that line, 'twill be an adequate cause of the motion of the Earth."[25] Like Cesalpino, then, White refused, as an Aristotelian, to confer *natural* circular or orbital motions on the earth. However, because he was convinced of the truth of heliocentric astronomy and that the earth really moved as the Copernican Theory required, he derived its motions by appeal to an external force.[26]

The explanations of Cesalpino and White represented a basic model for those few Aristotelians who sought an accommodation with the new geokinetic astronomy. While retaining the basic Aristotelian principle that the element earth could possess only one simple natural motion, which was downward and rectilinear, they were yet prepared to allow that *external forces* could cause the earth to move with one or more circular motions. Even those Aristotelians who disagreed with this approach could see its attractions. Raphael Aversa (1589–1657) conceded[27] that changes observed in the celestial region might well be saved by the assumption of a terrestrial motion, especially an axial rotation of the earth every twenty-four hours which could properly account for the same motions that many attribute to the heaven itself. But in truth, "every apparent local change around the celestial bodies really and truly happens to those bodies by a real and true motion; but no such change and motion occurs to the earth. This is the common sense of both wise and ordinary men."[28] Any observed motions that alter the relations between celestial bodies must be assumed to occur

---

[25] Ibid., Lesson XIV, 175, paragraphs 4–5. It is noteworthy that nowhere is the name of Copernicus mentioned.

[26] White's opinions and explanations of the earth's motions were presented earlier in his better known *De mundo dialogi tres* (Paris, 1642; see John L. Russell, "The Copernican System in Great Britain," in Jerzy Dobrzycki [ed.], *The Reception of Copernicus' Heliocentric Theory* [Dordrecht/Boston: D. Reidel, 1972], 222–223), a work that Thomas Hobbes saw fit to criticize systematically in a treatise that has only recently been published from a single manuscript at the Bibliothèque Nationale. For Hobbes's severe critique of White's explanations of the earth's motions, see *Thomas Hobbes: Thomas White's "De Mundo" Examined*, the Latin translated by Harold Whitmore Jones (London: Bradford University Press in association with Crosby Lockwood Staples, 1976), ch. 18, pp. 193–211. Hobbes's Latin text has been edited by Jean Jacquot and Harold Whitemore Jones, *Thomas Hobbes: Critique du "De Mundo" de Thomas White* (Paris: Librairie Philosophique J. Vrin, 1973). Not only did White provoke Hobbes's criticism, but his ideas were also repudiated by the papacy. "On 17 November 1661 the Holy Office condemned eight of his books explicitly (and implicitly all of his other writings, both past and future)" (see "White, Thomas," *Dictionary of Scientific Biography*, vol. 14, p. 301).

[27] Aversa, *Philosophia*, 2, question 34, sect. 5: 141, col. 2–142, col. 1.

[28] "Denique de facto omnis apparens mutatio localis circa corpora caelestia vere et realiter convenit ipsis corporibus caelestibus per verum et realem motum nullaque huiusmodi mutatio aut motus terrae convenit. Hic est communis tam Sapientum quam vulgarium hominum sensus." Ibid., 141, col. 2–142, col. 1.

in the heaven itself. Mobility is proper to the heaven and immobility to the earth, which lies in the middle or center of the world. Indeed on the principle that "to one simple body, [only] one motion is appropriate," the "appropriate" motion for the earth is downward and rectilinear. But Aversa conceded that this constraint would prove no obstacle "if circular motion were attributed to the earth from another extrinsic cause, or even from another motive power."[29] These were alternatives, however, that held little attraction for Aversa and most Aristotelians.

---

[29] ". . . uni autem corpori simplici unus motus competere debet. Sed hoc non obstaret, si motus circularis tribueretur terrae ab alia causa extrinseca sive etiam ab alia distincta virtute motiva." Ibid., 142, col. 1.

II. THE BASIC DEFENSE
OF ARISTOTELIAN COSMOLOGY

Although it is well to realize that the Aristotelian response was varied and could embrace radically divergent views, the most common reaction in the sixteenth and seventeenth centuries was one of unswerving support for the earth's centrality and immobility. But what form did that allegiance to Aristotle take? What were the specific arguments in defense of the traditional geocentric position? Were there significant disagreements among those who supported the earth's centrality and immobility? As the best means of conveying what may be characterized as the resistance to Copernicanism, I shall describe the defense of Aristotelian cosmology after the 1616 condemnation of Copernicus's *De revolutionibus orbium celestium*. The defense is represented by the works of certain authors who had some degree of familiarity with at least the major cosmological ideas and arguments in Copernicus's treatise. In a few cases, there was also an awareness of Galileo's defense of Copernicanism.

Six authors, who composed five relevant works over a thirty-year period between 1627 and 1657, will represent the scholastic reaction to Copernican claims for the earth's motions. Five of the six wrote primarily as natural philosophers considering traditional problems in Aristotle's physical treatises. Two of the five, Bonaventura Bellutus (ca. 1596–1676) and Bartholomew Mastrius (1602–1673), were Franciscan Conventuals and defenders of the philosophy of John Duns Scotus.[30] For twelve years, between 1628 and 1640, the two friars, who were described as "two minds in one soul, and one soul in two bodies,"[31] published a series of disputations on the logical and physical works of Aristotle. Of these treatises, the one on *De caelo*, which first appeared in 1640, is relevant to this study.[32] Two others were Jesuits; namely, Bartholomew Amicus (1562–1649), whose opinions were presented in a commentary on Aristotle's *De caelo* published in 1626,[33]

---

[30] On Mastrius and his relationship with Bellutus, see Bonaventure Crowley, O. F. M. Conv., "The Life and Works of Bartholomew Mastrius, O. F. M. Conv.," *Franciscan Studies*, 8 (1948): 97–152.

[31] Cited by Crowley, ibid., 117 from a work by Giovanni Franchini, who was Procurator General of the Order and was personally acquainted with Mastrius (Crowley, ibid., 98).

[32] See above, n. 22.

[33] *In Aristotelis libros De caelo et mundo dilucida textus explicatio et disputationes in quibus illustrium scholarum Averrois, D. Thomae, Scoti, et Nominalium sententiae expenduntur earumque tuendarum probabiliores modi afferuntur*. Auctore P. Bartolomaeo Amico, Societatis Jesu theologo. Tomus unicus. Naples: apud Secundinum Roncaliolum, 1626. This was the fifth in a series of seven volumes published by Amicus under the general title *In universam Aristotelis philosophiam notae et disputationes, quibus illustrium scholarum Averrois, D. Thomae, Scoti et Nominalium sententiae expenduntur earumque tuendarum probabiliores modi afferuntur* (Naples, 1623–1648).

and Melchior Cornaeus (1598–1665), professor of theology at the University of Würzburg (Herbipolis), whose views were incorporated into "a curriculum of Peripatetic philosophy as it is customarily covered in the schools at this time," in 1657.[34] Raphael Aversa (1589–1657), the fifth author, was a Carmelite priest and professor of theology at Rome, whose relevant discussions appear in the context of sixty questions ranging over the whole of Aristotle's physics, cosmology, and metaphysics.[35] The sixth author is Giovanni Baptista Riccioli (1598–1671), a famous Jesuit. Unlike the five authors already cited, who were not scientists properly speaking but natural philosophers in the medieval sense using problems in Aristotle's *De caelo* and *Physics* as the vehicle for their discussions, Riccioli was a technical astronomer and scientist, who considered the problem of the earth's immobility or mobility in his *New Almagest* (1651), a lengthy astronomical and cosmological treatise, the second part of which contains one of the most extensive discussions of the earth's status written during the sixteenth and seventeenth centuries.[36]

By the seventeenth century, the Copernican theory had caused a considerable weakening of support for the traditional Aristotelian-Ptolemaic astronomy and cosmology. If only by way of reaction, it led Tycho Brahe to devise his famous geoheliocentric system. Although Amicus and Aversa appear to have been defenders of Ptolemy's version of the traditional

---

[34] My translation is from the title of Cornaeus' treatise: *Curriculum philosophiae peripateticae, uti hoc tempore in scholis decurri solet. Multis figuris et curiositatibus e mathesi petitis, et ad physin reductis, illustratum. Autore R. P. Melchiore Cornaeo, Soc. Iesu, SS. Theologiae Doctore, eiusdemque ab Alma Universitate Herbipolensi Professore Ordinario* (Herbipoli: sumptibus et typis Eliae Michaelis Zinck, 1657). Brief biographical and bibliographical notices for Amicus and Cornaeus appear in Carlos Sommervogel S. J., *Bibliothèque de la Compagnie de Jésus,* Nouvelle édition (12 vols.; Brussels: Oscar Schepens; Paris: Alphonse Picard, 1890–1912), 1: cols. 279–280 (for Amicus, cited as Amici) and 2: cols. 1467–1471 (for Cornaeus).

[35] For the title of Aversa's work, see above, n. 22. By the unusual title of his work, *Philosophy United by Questions Embracing Metaphysics and Physics,* Aversa sought to distinguish it from logic and theology (see Reif, *Natural Philosophy in Some Early Seventeenth Century Scholastic Textbooks,* 11–12. The questions relevant to our interests appear in the second volume published in 1627. Little is known of Aversa, who was apparently born at San Severino (Salerno) and died at Rome in 1657, where, according to Charles H. Lohr ("Renaissance Latin Aristotle Commentaries: Authors A-B," *Studies in the Renaissance,* 21 [1974]: 253), he was not only professor of theology, but in "1623 rector of the college of the Order, there; praepositus generalis of the Order" and at some time "rejected the bishoprics of Nocera and Nardò." Reif (ibid., 11) connects his professorship in theology with San Severino rather than Rome. For a brief description of some of Aversa's ideas unconnected with the present investigation, see Lynn Thorndike, *A History of Magic and Experimental Science* (8 vols.; New York: Columbia University Press, 1923–58), 7: 393–96.

[36] *Almagestum novum astronomiam veterem novamque complectens observationibus aliorum, et propriis novisque theorematibus, problematibus ac tabulis promotam; in tres tomos distributam quorum argumentum sequens pagina explicabit. Auctore P. Ioanne Baptista Ricciolo, Societatis Iesu Ferrariensi* (Bologna: ex typographia Haeredis Victorij Benatij, 1651). Although the title page indicates a three-volume work, only one volume appeared, which was divided into two parts, each separately paginated. All the material relevant to our study apears in the second part. For a bibliography of Riccioli's scientific works, see Luigi Campadelli, "Riccioli, Giambattista," in *Dictionary of Scientific Biography,* 11: 411–12.

geocentric cosmology, Melchior Cornaeus supported Tycho Brahe's formulation.[37] Moreover because he proclaimed on the title page that his book contained "the curriculum of Peripatetic philosophy as it is customarily covered in the schools at this time," Cornaeus's support for the Tychonic system may indicate that the latter was taught and supported in many Jesuit schools. Riccioli also sided with Tycho, but formulated his own versions of that system to counter the Copernicans.[38] The new astronomical systems also influenced Mastrius and Bellutus, who adopted what was known as the Capellan system, whereby they assumed the configuration attributed by Copernicus to Martianus Capella, namely that Mercury and Venus move around the sun, while the latter and all the other celestial bodies rotated around a stationary earth.[39]

Despite these major differences, our authors were all agreed on a geocentric universe. They rejected the daily and annual motions of the earth and were committed to a defense of the earth's centrality and immobility. Since the primary purpose of this study is to describe and assess the battery of scholastic arguments presented in favor of the earth's centrality and immobility in the three or four decades following the condemnation of the Copernican system, the astronomical and cosmological differences between our authors is much less significant than is their basic agreement on a geocentric universe with a stationary earth.

Any scholar who wishes to assess seventeenth century scholastic arguments in favor of the earth's centrality and immobility confronts a dilemma. Did the condemnation of Copernicus, Diego de Zuniga, and Paolo Foscarini in 1616 by the Congregation of Cardinals[40] and of Galileo

---

[37] After describing the Ptolemaic astronomy of the Jesuit astronomer Christopher Clavius and the heliocentric system of Copernicus, both of which he criticized, Cornaeus declares that Tycho Brahe constructed the best system of all (*Curriculum philosophiae peripateticae, Disputatio III, De elementis,* Dub. 9, 525–28).

[38] *Almagestum Novum,* pars prior, bk. 3 (De sole), p. 103, col. 1. According to Christine Jones, *The Geoheliocentric Planetary System,* 176–77, Riccioli characterized one system as "semi-Ptolemaic," which is really what was known as the Capellan system (see below and n. 39), and the other, to which he apparently attached greater significance, was called "semi-Tychonic" wherein Jupiter and Saturn had earth-centered orbits and Mars moved about the sun as center.

[39] Mastrius and Bellutus, *De coelo,* 489, cols. 1 and 2. Lynn Thorndike mentions this in his *History of Magic and Experimental Science,* 7: 468 (he used the edition of Venice 1688); see also Jones, *The Geoheliocentric Planetary System,* 301–302. Capella described the system named after him in his famous *Marriage of Philology and Mercury.* For Copernicus's reference to Martianus Capella, see Rosen (tr.), *Nicholas Copernicus On the Revolutions,* bk. 1, ch. 10, 20, lines 5–12.

[40] Amicus, Aversa, and Riccioli mention the condemnation of 1616. See Amicus, *De caelo,* Tract 5, question 6 ("On the Motion of the Heavens"), Doubt 1 ("Whether the starry heavens are moved around an immobile earth, or the contrary"), 291, col. 1, where Amicus actually quotes the text of the condemnation; Aversa, *Philosophia,* question 31, section 2 in 2: 5, col. 2. Riccioli not only published the text of the condemnation of 1616, but also the passages from the *De revolutionibus* that were judged offensive (see *Almagestum novum,* pars posterior, 496–97). For a translation of the condemnation of 1616, see Jerome J. Langford, *Galileo, Science and the Church,* rev. ed., foreword by Stillman Drake (Ann Arbor: The University of Michigan Press, 1971; first published in 1966), 97–98.

in 1633 compel the falsification of arguments by those scholastic theologians who may have been dubious about the traditional Aristotelian position and open minded about, and perhaps even receptive to, Copernicus's claims about the earth's mobility? Since all the scholastic authors considered here wrote after 1616, and a few after 1633, the question is obviously relevant. It becomes even more relevant when one realizes that three of the six authors discussed here were Jesuits. For it is well known that many Jesuits found Copernican astronomy unobjectionable[41] and perhaps even better founded than any other available system. But the aftermath of the condemnation of Galileo in 1633 saw Jesuit scientists pressured by the Roman Church "to reinforce the Decree of 1633 by publishing books on the controversy themselves emphasizing the religious aspect." "There was," consequently, "a spate of such books by Jesuit writers, in which ostentatious reference to the decision of the Church was made."[42] Both Riccioli and Cornaeus, who wrote after 1633, cited it, with Riccioli actually including the text.[43]

Now if those scholastics who believed that the Copernican theory was more appropriate, or at least no worse, than the various contemporary geocentric systems were also compelled to repudiate the Copernican system for theological reasons, might this not have affected the sincerity of their arguments in favor of a motionless and central earth? Indeed Riccioli, of whom Delambre would say "without his robe he would be Copernican,"[44] may have embodied this very dilemma. Although he argued vigorously against the Copernican system, and even characterized as unanswerable some of his own arguments for terrestrial immobility, Riccioli also rebutted some arguments for terrestrial immobility by invoking counterarguments from "the Copernicans," which seemingly left the earth's immobility in doubt.[45]

---

[41] See Jones, *The Geoheliocentric Planetary System*, 285.

[42] Ibid., 281.

[43] For Riccioli, see *Almagestum novum*, pars posterior, 497–99. Riccioli also included the text of Galileo's abjuration (ibid., 499–500; for a translation of Galileo's sentence and abjuration, see Giorgio de Santillana, *The Crime of Galileo* [Chicago: The University of Chicago Press, 1955], 306–310, 312–13). Not only did Cornaeus mention the condemnation of Galileo in 1633, but he even included the fact that during the first three years of confinement, Galileo was required to recite the seven penitential Psalms once every week. Also mentioned was the 1616 condemnation of Copernicus's *De revolutionibus* (see *Curriculum philosophiae peripateticae*, 536–37). Mastrius and Bellutus (*De coelo*, 562, col. 1, par. 112) mention only that the opinion opposed to the earth's centrality and immobility "was damned by the Sacred Congregation of Cardinals and assigned to the index of books" ("hinc opposita opinio damnatur a Sacra Card. Congreg. ad indicem librorum deputata. . . .").

[44] Translated by Jones, *The Geoheliocentric Planetary System*, 286 from J. B. Delambre, *Histoire de l'Astronomie Moderne*, 2 vols. (Paris, 1821), 2:279.

[45] Indeed in a summary of thirty-eight arguments in support of the earth's immobility, Riccioli, in the thirtieth argument (*Almagestum novum*, pars posterior, Bk. IX, section IV, ch. 24, 475, col. 1), presented the common opinion that if the earth rotated, we ought to perceive it. Since we do not, one may infer the earth's immobility. But Riccioli, citing an earlier fuller discussion (ibid., ch. 22, number 7, 433) insisted that "there is no necessity for this sensation" ("At nullam revera esse necessitatem sensationis huius docuimus cap. 22, num. 7"). In view

If Riccioli was a secret Copernican[46] did he and others like him subtly alter arguments and surreptitiously make the earth's mobility appear more plausible than its immobility? No evidence for such an interpretation exists. Indeed one might even argue that by conceding that certain arguments were indecisive for either side, but insisting on the validity of certain others, Riccioli may have proved more effective in advancing the cause of terrestrial immobility than if he had inflexibly sought to repudiate every pro-Copernican argument. But even if Riccioli was a secret Copernican and subtly attempted to undermine the anti-Copernican position whenever feasible, the arguments he presented, many of which were traditional and well known, must nevertheless be considered at face value and be compared to similar or identical arguments by others. While the motives and innermost convictions of authors like Riccioli are important where they can be discerned, they do not affect our treatment. Whatever secret reservations he and our other authors may have had may never be known. Until they are, our task is straightforward: to examine the arguments, compare them to others, and assess them as if they were proposed sincerely.

The discussions of our authors mark the culmination of more than 400 years of geocentric cosmology in western Europe. Their extensive arguments represent the cumulative wisdom of a long tradition that is everywhere evident by the sources and authorities they cite, which may be conveniently divided into three categories. The first includes those ancient Greek and Latin treatises that became available during the late fifteenth and sixteenth centuries (described above) and which significantly supplemented what had been known during the Middle Ages. A second major source derived from the late Middle Ages by way of certain treatises that were published and made available to late sixteenth and seventeenth century scholastic authors. Of particular significance within this group were the *Questions on "De celo et mundo"* of Albert of Saxony (ca. 1316–1390),[47] the *Questions on the "Sphere"* of Sacrobosco of Pierre d'Ailly (1350–1420),[48] and the *Com-*

---

of the obvious importance of such an argument for the anti-Copernican position, this appears a rather remarkable concession. In the twenty-ninth argument (ibid., 475, col. 1), Riccioli declares that a rotating earth ought to cause the collapse and ruin of buildings and the projection from the earth's surface of things not securely attached to it. He then concludes the argument with a rebuttal from the Copernicans when he says that "the reply is that there is no such danger in the Copernican hypothesis, as is obvious from the statements in ch. 22, number 6."

[46] With regard to Riccioli's two astronomical schemes mentioned above, Jones observes (*The Geoheliocentric Planetary System,* 286) that Riccioli "made no attempt to proselytise, or even to defend these schemes with any enthusiasm."

[47] For the numerous editions and for the specific edition cited in this article, see above, n. 6. Albert of Saxony is an important link because he often reflected, and occasionally even repeated verbatim, arguments from John Buridan's (ca. 1295–ca. 1358) *Questions on De caelo,* which contained some of the best medieval argumentation on the earth's centrality and immobility but was left unpublished until 1942 (see Ernest A. Moody, *Iohannis Buridani Quaestiones super libris quattuor De caelo et mundo* [Cambridge, Mass.: The Mediaeval Academy of America, 1942]).

[48] Where necessary, I shall cite the edition of 1515 (*Habes lector Johannis de Sacro Busto Sphere textum una cum additionibus non aspernandis Petri Cirvelli. D [a vero tamen textu apparenter*

mentary on the Second Book of the Sentences of Peter Lombard and the *Questions on "De celo"* of John Major (1469–1550).[49] From these[50] and other medieval predecessors, Aristotelian scholastic authors received the context, the form, and often the content of the disputes about the earth's centrality and immobility, especially as these had been shaped in the fourteenth century. As the third group, we must take cognizance of the immediate scholastic predecessors of our authors, especially those who wrote during the late sixteenth and early seventeenth centuries. The most prominent members of this group were Christopher Clavius (1537–1612), whose *Commentary on the "Sphere" of Sacrobosco*[51] was widely read, and the Jesuits at the University of Coimbra, known collectively as *Conimbricenses*,[52] who wrote

distinctis] *cum ipsiusmet sublimi et luculentissima expositione aliquot figuris noviter adiunctis decorata; intersertis preterea questionibus Domini Petri de Alliaco* [Paris: J. Parvo, 1515]) which contains the text of Sacrobosco's *Sphere* along with Pedro Cirvelo's (1470–1554) *Commentary on the Sphere* and Pierre d'Ailly's *Questions on the Sphere,* both of which are interspersed through Sacrobosco's text. The edition of 1515 is perhaps the same as the editions of 1498/99 and 1508 also published by Jean Petit [i.e. J. Parvo] with the title *Uberrimum sphere mundi commentum intersertis etiam domini Petri de Aliaco.*

[49] For the Commentary on the Sentences, see *Editio secunda Johannis Maioris doctoris Parisiensis in secundum librum Sententiarum nunquam antea impressa* (Paris, 1519), Distinction 14, question 12 (on the center of magnitude and gravity of the earth), fols. 84v–85v (?). Major's *Questions on De celo* appears in his *Octo libri Physicorum cum naturali philosophia atque Metaphysica Johannis Maioris Hadingtonani theologi Parisiensis* (Paris: Jean Petit, 1526). Although the *Questions on De celo* is not mentioned on the title page (nor are his questions on the *Meteorologia* and *De generatione et corruptione*), it appears, without any break, immediately after the *Questions on the Physics* at sig.k iiii recto (the volume is unpaginated). There are no questions on the third book. An index of questions for all the treatises included in the volume appears at the end of the book.

Although Major lived through the first half of the sixteenth century, he is classified here as "medieval" because his Aristotelian scholastic commentaries were in the fourteenth century Parisian tradition, a tradition he helped revive at Paris during the first few decades of the sixteenth century. Because he was often cited by scholastics in the late sixteenth and seventeenth centuries, he must be regarded as a significant source of late medieval ideas in the seventeenth century.

[50] Among these three authors, we can also establish a pattern of influence. There can be little doubt that Albert of Saxony's *Questions on "De celo"* influenced Pierre d'Ailly's *Questions on the "Sphere" of Sacrobosco* (on this, see Pierre Duhem, *Le Système du monde,* vol. 9, p. 231 and below in this article) and that both influenced John Major's discussions on the earth in the latter's *Commentary on the Second Book of the Sentences of Peter Lombard.*

In the category of "other" authors, we must mention the works of Thomas Aquinas and Duns Scotus as turning up in the discussions, as indeed also the commentaries on the *Physics* and *De caelo* of Albertus Magnus. Other names might also be mentioned (for example, John of Jandun, Durandus, Aegidius Romanus, etc.) but none are as important as the first three who disseminated significant ideas formulated in the fourteenth century, ideas that were often derived ultimately from John Buridan and Nicole Oresme.

[51] *Christophori Clavii Bambergensis ex Societate Iesu in Sphaeram Iohannis de Sacro Bosco Commentarius* (Rome: Apud Victorium Helianum, 1570). According to W. G. L. Randles (*De la terre plate au globe terrestre, une mutation épistémologique rapide 1480–1520* [Paris: Librairie Armand Colin, 1980], 48, 95), there were eight editions in the sixteenth century and nine more in the seventeenth century, the last in 1618. I shall cite the fourth edition of Lyon, 1593, which bears the same title as the edition of 1570 (for further details, see Sommervogel, *Bibliothèque de la Compagnie de Jésus,* Vol. 2, cols. 1212–1213, where a translation into Chinese by Father Matteo Ricci is also mentioned).

[52] Of the numerous commentaries on Aristotle's natural and logical works, I shall cite only the following edition: *Commentarii Collegii Conimbricensis Societatis Iesu In quatuor libros De*

popular commentaries on the works of Aristotle among which the commentary on *De caelo* was of particular importance.[53] As representatives of scholastic opinion in the seventeenth century, our authors were, to different extents, also knowledgeable about the new cosmological ideas embodied in the works of Copernicus, Tycho Brahe, Kepler, Galileo, Gilbert, and other non-scholastics.[54] Indeed the arguments in favor of the earth's motions found in these works compelled scholastics to respond and thus devise new arguments that were added to the store of traditional objections to the earth's motion.

The defense of the Aristotelian geocentric world view in the seventeenth century was largely the work of scholastic theologians, rather than Aristotelian humanists or secular natural philosophers. The six authors on whom our attention will be focused are thus excellent representatives of seventeenth century scholastic opposition to the Copernican theory, especially since Amicus, Riccioli, and Cornaeus were members of the Jesuit Order, which, after the condemnation of 1616, led the opposition against the new cosmology. Because commentators on Aristotle's *De caelo* and Sacrobosco's *Sphere* regularly considered the problem of the earth's centrality independently of its alleged immobility—indeed, as we saw with Cesalpino, it was possible to accept the former without the latter—we shall also address them separately, taking up centrality first, as was the custom.

The scholastic defense of the earth's centrality and immobility can be, and often was, divided into three parts: astronomical, physical, and Scrip-

---

coelo Aristotelis Stagiritae (2d ed.; Lyon: ex officina Iuntarum, 1598). Although this commentary was actually composed by Emmanuel de Goes, S. J. (1542–1597) and first published in 1592, the works of the Coimbra Jesuits were issued anonymously and usually referred to by their collective title "Conimbricenses," a practice that I shall also follow (under this same title, see also, Charles H. Lohr, "Renaissance Latin Aristotle Commentaries: Authors C," *Renaissance Quarterly* 28, Nr. 4 [1975]: 717–719).

[53] Other authors in this third group of immediate scholastic predecessors are Scipio Claramontius, Johannes Cottunius, Johannes Magirus, Bartholomew Keckerman, Roderigo de Arriaga, Ruvius, and Tanner.

The scholastic theologians who defended the geocentric world view in the seventeenth century also occasionally cited sixteenth-century Aristotelians who were not scholastics but, for lack of a better term, may be classified as humanists. In this group, we may mention the following commentaries on *De caelo* by Jacques Le Fèvre d'Etaples (*Totius naturalis philosophiae Aristotelis paraphrases per Iacobum Stapulensem . . . Introductio in libros Physicorum. . . . Quatuor De caelo et mundo completorum paraphrasis . . .* [Friburgi Brisgoiae, Excudebat I. Faber Emmeus, 1540), Lucillus Philaltheus (pseudonym for Lucilio Maggi) (*Lucilli Philalthaei, philosophiae medicinaeque professoris publici, In IIII libros Aristotelis De caelo et mundo commentarii . . .* [Venice: apud Vincentium Valgrisium, 1565]), and Polus Lauredanus (*In Aristotelis De coelo libros quatuor Poli Lauredani Patritij Veneti Commentaria . . .* [Venice: apud Ioannem Baptismam Ciottum Senensem, 1598]). Works containing peripatetic questions also frequently included relevant discussions, as, for example, those by Andrea Cesalpino (cited above, n. 18) and by Francesco Patrizi (*Francisci Patricii Discussionum peripateticarum tomi IV* [Basileae: ad Perneam lecythum, 1581]).

[54] Riccioli, who cites all of these authors and many more, had apparently read them directly. The range of Riccioli's quotations and references to both scholastic and nonscholastic literature is truly remarkable.

tural.[55] Because even some of its opponents conceded the astronomical soundness of the Copernican system, the attacks against it tended to emphasize the physical and Scriptural, rather than astronomical, objections to the earth's motions.[56] The contexts within which the Copernican theory were discussed varied. Amicus described and refuted the theory in the first part of a question "On the Celestial Motions" (*De motu caelorum*), where, in the first of a series of doubts he inquires "whether the starry heavens are moved around an unmoved earth, or whether the contrary [is true]";[57] Aversa concentrated his attack in a question on the order and arrangement of the various parts of the world,[58] which formed part of a larger section dealing with the celestial and terrestrial bodies; Melchior Cornaeus, who included much on technical astronomy, mathematics, and optics, chose, by contrast, to present the Copernican theory and its refutation in a section on the elements (*De elementis*) within that part of his lengthy treatise devoted to the explication of Aristotle's *De caelo*.[59] Like Cornaeus, Mastrius and Bellutus considered the Copernican theory in their treatise on *De caelo* within the overall framework of a discussion on the elements and within a further subdivision concerned with earth and water and their mutual relations. It was within this context, that they finally inquired "whether this globe composed of earth and water could be moved circularly."[60]

The authors just mentioned considered the problem of the earth's mobility or immobility within a context of traditional problems posed in commentaries on Aristotle's *De caelo*. Only Riccioli differed from this pattern, since his treatise, as the title suggests, is astronomical and cosmological. It is far removed from the format of typical commentaries on *De caelo*, although it could hardly avoid problems that also were raised in the latter, of which the earth's centrality and immobility were of special importance.

---

[55] In a section on the order and position of the parts of the world, Raphael Aversa considered the three parts separately and so labeled them. See his *Philosophia,* Question 31 (*De mundo*), section 2, pp. 5, col. 1–6, col. 1.

[56] See Jones, *The Geoheliocentric Planetary System,* 285.

[57] Bartholomeus Amicus, *De caelo et mundo,* Tract 5 ("On the Properties and Perfections of the Heavens" [*De proprietatibus et perfectionibus coelorum*], Question 6 ("On the Motion of the Heavens" [*De motu caelorum*], Doubt 1 (Dubitatio 1: *An caeli stellati moveantur circa terram immotam, an vero contra*), 288, col. 2–292, col. 1.

[58] Raphael Aversa, *Philosophia,* question 31 (*De mundo*), second section, "In what order and position are the special parts of the world arranged?" (*Quo ordine et situ dispositae sint partes praecipuae mundi*), 2: 4–6. In Question 36, fifth section (ibid., 224, col. 1–231, col. 1), which is concerned with whether the earth is in the center of the world, Aversa briefly mentions Copernicus and where the latter located the earth within the world system and the elementary world (see especially 224, cols. 1–2).

[59] Melchior Cornaeus, *Curriculum philosophiae peripateticae,* Part III, *Libri IV De coelo et mundo,* Disputation III: On the Elements (*De elementis*), question 2 ("On Certain Affections of the Elements" [*De quibusdam affectionibus elementorum*], 516–538, but especially 525–538.

[60] Mastrius and Bellutus, *De coelo,* Disputatio quarta ("De elementis in particulari, ubi etiam de Meteoris," p. 538), quaestio IV ("De aqua et terra eorumque Meteoris," p. 558, col. 2), Articulus III ("An globus iste ex terra et aqua compactus moveatur circulariter," 562, col. 1).

Riccioli considered the problem of the earth's motion in two sections of Book IX of the *New Almagest*, which treated of the "System of the World" (*De mundi systemate*). In section IV of that book, he considered the "System of a Moved Earth," where, in 117 double columned pages (the first eighteen chapters of section IV), he presented and refuted virtually the whole range of arguments offered in favor of the earth's different possible motions.[61] In the next seventy pages (chapters 19–35), Riccioli argued specifically for the earth's immobility.[62] His significance for this study cannot be overestimated. Not only did Riccioli compose what may well be the most extensive and detailed defense of geocentric astronomy and cosmology ever written, but he was familiar with an amazing array of authors and treatises for and against the traditional astronomy, including the relevant works of Amicus, Mastrius, and Bellutus. Moreover, he was probably the last scientist of any standing to defend the centrality and immobility of the earth in a major astronomical treatise.

---

[61] Riccioli, *Almagestum novum*, pars posterior, 290–407. Within these pages, Riccioli included forty-nine specific arguments, along with thirteen classes of experiments.

[62] Ibid., 408–479. Here Riccioli musters at least seventy-seven arguments for the earth's immobility. He then presents summaries of all earlier arguments: twenty for the earth's daily motion (466–469); twenty-nine for the annual motion of the earth (469–472; the total for the earth's motions is thus forty-nine); thirty-eight against the simultaneous daily and annual motions of the earth (472–475) and thirty-nine against the annual motion of the earth (475–477; the total for the earth's immobility is thus seventy-seven). Earlier in the *Almagestum novum*, pars prior, bk. II, 49–52, Riccioli briefly described different opinions about the earth's centrality and immobility, but informs the reader that the major discussion will appear in bk. IX.

III. THE EARTH'S CENTRALITY

A. THE THREE CENTERS

As heirs to a long, evolving Christian Aristotelian tradition, our authors were convinced on physical and Scriptural, if not astronomical, grounds that our earth is an elemental, spherically shaped, immobile body whose center coincides with the center of a spherical universe. Modern scholarship has depicted these convictions oversimplistically and thus distorted the basic Aristotelian opposition to the Copernican theory. Such distortions may have resulted from the unqualified use of key terms like "center," "earth," "spherical," and "immobile." Centuries of discussion had produced different senses and significant qualifications for each of these interrelated terms.

Although numerous arguments were advanced for belief in the earth's centrality, just what lay at the center of the world and what it was the center of were by no means clear. The difficulties and confusions are traceable to a distinction already made in the fourteenth century by John Buridan between the earth's center of gravity and its center of magnitude, a distinction in which Archimedean ideas about centers of gravity "began . . . to play a role in the mechanics of large bodies."[63] Transmitted to the seventeenth century in the works of Albert of Saxony, Pierre d'Ailly, John Major, and others, the medieval versions of the definitions of the three centers—geometric center, center of gravity, and center of magnitude— were frequently cited. They are found, for example, in Aversa's *Philosophia*, where the center of the universe is defined as that "from which all lines drawn to the circumference of the lunar heaven [or sphere] are equal, as are the lines drawn to the extreme and supreme circumference of the whole heaven."[64] The center of a body, and therefore of the earth itself, however, is conceived in two ways:

---

[63] For Buridan's discussion in his *Questions on De caelo*, Bk. 2, question 22 ("Whether the earth always is at rest in the center of the universe"), see the translation in Marshall Clagett, *The Science of Mechanics in the Middle Ages* (Madison, Wis.: University of Wisconsin Press, 1959), 594–599; for the quotation above, which is by Clagett, see p. 591. For lack of evidence, Clagett concludes that Archimedes did not directly influence fourteenth-century discussions on the earth's center of gravity. Relevant discussions on the earth's center of gravity appear in Pierre Duhem, *Le Système du monde. Histoire des doctrines cosmologiques de Platon à Copernic* (10 vols.; Paris: Hermann, 1913–1959), 9: 293–323; Henri Hugonnard-Roche, *L'Oeuvre astronomique de Themon Juif Maître Parisien du xive siècle* (Geneva: Librairie Droz; Paris: Librairie Minard, 1973), 76–86, which summarizes question 4 ("Whether the earth rests naturally in the middle of the world") of Themon Judaeus's *Questions on the "Sphere" of Sacrobosco*; and Thomas Goldstein, "The Renaissance Concept of the Earth in Its Influence upon Copernicus," *Terrae Incognitae*, The Annals of the Society for the History of Discoveries, 4 (1972): 32–33.

[64] "Pro qua re considerandum est punctum medium ipsius mundi, quod dicitur centrum, et est illud a quo omnes lineae ductae ad circumferentiam caeli Lunae sunt aequales inter

one is called the *center of gravity* and the other of *magnitude*. The center of gravity is explained as the mid-point of any line in which, if the body were divided, each half would be of equal weight. The center of magnitude is explained as the point from which all lines drawn to the circumference are equal; or [it is] the mid-point of any line by which, if the body were divided, both halves would be of equal magnitude.[65]

From these definitions, the relationship between the centers of magnitude and gravity had to be determined. It was generally assumed that if a given sphere were of homogeneous composition, the two centers would coincide. But if its composition were heterogeneous (*heterogeneus*, to use Cornaeus's term), or non-uniform (*difformis*, as Clavius put it), and it was composed of, say, one hemisphere of lead and one of wood, the two centers would differ, with the center of magnitude coincident with the geometric center, while the center of gravity would be located somewhere within the lead hemisphere.[66] To be sure, the distinction described here was in no way

se, et pariter lineae ductae usque ad circumferentiam extremam et supremam totius caeli sunt aequales inter se." Aversa, *Philosophia,* vol. 2, question 36, fifth section, p. 225, col. 1. Although Cornaeus omitted this one, Amicus defines the center of the universe as "that which is equidistant from all parts of the circumference of the universe" ("Nam illud quod aequaliter distat ab omnibus partibus circumferentie universi est centrum universi." *De coelo,* 582, col. 1). For the definition by the Conimbricenses, who extend the equal lines to the *primum mobile,* see their *Commentary on De Caelo,* bk. 2, question 3, article 1, 381.

[65] "Ubi rursus distingui solet duplex centrum alicuius corporis, unum dicitur centrum gravitatis et aliud magnitudinis. Centrum gravitatis explicatur punctum medium cuiusdam lineae, in qua si dividatur corpus, utraque medietas aequaliter ponderabit. Centrum magnitudinis explicatur punctum a quo omnes lineae ductae ad circumferentiam sunt aequales, sive punctum medium cuiusdam lineae, in qua si dividatur corpus, ambae medietates sint aequalis magnitudinis." Aversa, ibid. The italics are mine. Both definitions are strikingly similar to those offered by Pierre d'Ailly (*Questions on the "Sphere" of Sacrobosco,* question 5, fols. 27v, col. 2–28r, col. 1 of the edition cited above (n. 48) and John Major (*In secundum librum Sententiarum* [Paris, 1519], Distinction 14, question 12, fol. 84v, col. 1). In the sixteenth century, the Conimbricenses adopted substantially the same definitions (*De coelo,* p. 381). Among our authors, Mastrius and Bellutus, *De coelo,* question 4, article 2 ("An iste globus ex terra et aqua constitutus habeat idem centrum cum universo, ubi de motu trepidationis"), p. 560, col. 2, offer definitions that are quite similar to Aversa's, whom they cite frequently in this section. Although Amicus's definition of center of magnitude (see *De coelo,* 584, col. 1) is virtually identical with Aversa's and that of Cornaeus is similar (*Curriculum philosophiae peripateticae,* 519, Nota 1), their definition of center of gravity drew upon the first of two definitions of center of gravity (the second was drawn from the *Collectio* of Pappus of Alexandria) provided by Christopher Clavius who declared that "the center of gravity of any body is that point which always tends perpendicularly toward the center of the whole universe, however much and however often the body is suspended provided that it hangs freely" (*In Sphaeram Ioannis de Sacro Bosco Commentarius* [Lyon, 1593], 136). Cornaeus's definition (*Curriculum philosophiae peripateticae,* 519) reads much like Clavius's whereas Amicus's differs considerably and is given, without mention of suspended bodies, within a syllogism in support of the claim that the "center of gravity and [the center] of the universe are the same" (for the definition, see *De coelo,* 582, col. 1).

[66] Discussions that follow the general description given above appear in Aversa, *Philosophia,* 225, col. 1; Mastrius and Bellutus, *De coelo,* 560, col. 2, par. 105; Cornaeus, *Curriculum philosophiae peripateticae,* 519, Notandum 3; and Clavius, *Commentary On the "Sphere" of Sacrobosco,* 136–137. Although Amicus did not formally express the distinction between centers of gravity and magnitude, he assumed it. The distinction was already made in the fourteenth century by John Buridan (*Quaestiones super libris quattuor De caelo et mundo,* ed. E. A. Moody [Cambridge, Mass.: The Mediaeval Academy of America, 1942], Bk. 2, question

controversial or unusual. Problems arose, however, when it had to be applied to a real, physical earth. To know whether the earth's centers of gravity and magnitude coincided, or even if the earth had a center of magnitude, it was essential to have in mind a firm concept of the "true" nature of the earth. Was it a perfect square? Was it composed of the element earth alone or was water, as found in the vast seas and oceans, part of it?

B. THE TERRAQUEOUS SPHERE

In his famous *Commentary on the "Sphere" of Sacrobosco,* Christopher Clavius explained that although Sacrobosco located the earth in the center of the firmament, we should understand "the earth simultaneously with water. For although the author speaks expressly of the earth alone, the same arguments have the same force with respect to the whole aggregate of earth and water."[67] Although Sacrobosco did not himself conceive of the earth in the manner described by Clavius, the treatment of earth and water as a single aggregate was begun in the fourteenth century, when it was enunciated, and perhaps adopted, by Albert of Saxony, proclaimed unambiguously by Pierre d'Ailly in the late fourteenth or early fifteenth century, and again declared by John Major in the sixteenth century. Not until the early sixteenth century, however, and then only outside the scholastic tradition, was this aggregate conceived as a single sphere, an interpretation assumed by Copernicus in Bk. 1, ch. 3 of *On the Revolutions,* where he explains "How Earth Forms a Single Sphere with Water."[68] This "terraqueous sphere," as it has been aptly called,[69] would also find favor

---

7 ["Whether the whole earth is habitable"] 159). Albert of Saxony repeated the essence of it in his *Questions on "De celo",* Bk. 2, question 23, fol. 116v, col. 2, where he explains that the centers of magnitude and gravity differ in "difformly heavy bodies" but are one and the same in uniformly heavy bodies (in his *Questions on the Physics,* Bk. 4, question 5, fol. 46r, col. 2 [in the edition cited above, n. 6], Albert applied the distinction to the earth arguing that if the earth were of uniform heaviness, which it is not [for reasons that will be mentioned below], its centers of gravity and magnitude would be identical). Influenced perhaps by Albert of Saxony, the same distinction was made by Pierre d'Ailly, 14 *Questions on the "Sphere" of Sacrobosco,* question 5, fol. 28r, col. 1, and John Major, *In secundum librum Sententiarum,* Distinction 14, question 12, fol. 84v, col. 1.

[67] Clavius, *In Sphaeram Ioannis de Sacro Bosco Commentarius* (4th ed.; Lyon, 1593), p. 151. For the passage in Thorndike's edition and translation of Sacrobosco's *Sphere,* see pp. 84 (Latin) and 122 (English).

[68] See Edward Rosen (tr.), *Nicholas Copernicus On the Revolutions,* 9.

[69] The concept of the terraqueous sphere was in existence more than one hundred years before that term was actually employed in 1643 by the French Jesuit, Georges Fournier (*Hydrographie*) and repeated in 1651 by Riccioli (*Almagestum Novum*), who used the term *terraqueum.* See W. G. L. Randles, *De la terre plate au globe terrestre* (Paris: Librairie Armand Colin, 1980), p. 63. On pages 41–64, Randles presents a fascinating account of the development of the terraqueous sphere; for a much longer earlier account, though still of great importance, see Pierre Duhem, *Le Système du monde,* Vol. 9, chs. 16 ("L'équilibre de la terre et des mers. I. Les anciennes théories), 79–170 and 17 ("L'équilibre de la terre et des mers. II. La théorie Parisienne"), 171–235. For another excellent account, which focuses on Copernicus, but also describes the medieval contributions and Copernicus's departures therefrom, see Thomas

in scholastic cosmology and it is the opinion adopted in the seventeenth century by Amicus, Aversa, and Cornaeus.

Without emphasis or elaboration, Aristotle had arranged the four elements in a series of concentric spheres with the earth at the center of the universe surrounded by the spheres of water, air, and fire in that order.[70] Aristotle was, however, aware that nature did not fully conform to his schema since dry land extended above the waters and fire was visible on the earth's surface.[71] But the relations between the spheres of earth and water, on which Aristotle had provided little guidance, posed serious problems in the Middle Ages. In John of Sacrobosco's enormously popular treatise *On the Sphere*, the four sphere Aristotelian scheme is presented as the true picture of the sublunar world with the important proviso that dry land exists for animate creatures and thus prevents the sphere of water from completely surrounding the earth.[72] Although Sacrobosco offered no explanation for the earth's emergence above the waters, a Biblical interpretation was frequently invoked. After surrounding the earth with waters on the second day of creation, God, on the third day, commanded that "the waters under the heaven be gathered together into one place" and that "dry land appear."[73] With the waters gathered together upon divine command, one could imagine—as Paul of Burgos (ca. 1350–1435) did—that God had lowered the sphere of water and thereby separated the

---

Goldstein, "The Renaissance Concept of the Earth in its Influence upon Copernicus," *Terrae Incognitae* 4 (1972): 19–51.

No Latin equivalent of the term "terraqueous" was used by Copernicus or the scholastic authors whom we shall discuss below. The term was, however, used by Rosen in his comment on Copernicus's title for Bk. 1, ch. 3, about which Rosen declares (*Nicholas Copernicus On the Revolutions*, p. 345, note to P.9:17): "In his *Geography*, a work cited by Copernicus here in I, 3, Ptolemy asserted the unity of the terraqueous sphere: 'From the mathematical disciplines we obtain the proposition that the continuous surface of land and water, taken as a whole, is spherical' (I, 2, 7). It had been shown by Archimedes that the 'surface of any fluid at rest is spherical and has the same center as that of the earth' (*Floating Bodies*, I, 2)." As will be seen below, the Portuguese discoveries of land in the southern hemisphere played a role in gaining acceptance for the concept of a single sphere of earth and water. Since Copernicus cites those discoveries in Bk. 1, ch. 3, they probably served to confirm Ptolemy's assertion.

[70] See Aristotle, *Meteorology* 2.2.354b.23ff.

[71] See the comments of H. D. P. Lee in his translation of Aristotle's *Meteorologica* (Loeb Classical Library; 2nd ed.; London: William Heinemann Ltd; Cambridge, Mass.: Harvard University Press, 1962; first printed, 1952), 24–27. For a useful description of Aristotle's account of the relationship of the spheres of earth and water, see Goldstein, "The Renaissance Concept of the Earth," *Terrae Incognitae*, 4: 26–29.

[72] See Thorndike, *The "Sphere" of Sacrobosco and Its Commentators*, 78–79 (Latin); 119 (English) and Goldstein, ibid., 30–31.

[73] Gen.1.9; see also Pss. 103 (Vulgate) and Randles, *De la terre plate*, 11. According to Mastrius and Bellutus (*De coelo*, p. 560, col. 2–561, col. 1, par. 107), God placed the earth at the center of the world and surrounded it by waters. Under these pristine conditions, the earth was a homogeneous body in which every equal part was of equal weight. The earth's centers of gravity and magnitude coincided with the center of the universe. The gathering of the waters, however, caused a considerable change. In the resultant non-homogeneous terraqueous, or terrestrial, globe (ibid., p. 561, col. 1, par. 108), the three centers were no longer mathematically identical, although physically—that is, approximately— they could be considered in the same place (for the reason, see below, n. 106).

latter's center of gravity from the earth's,[74] an act that would leave all of the earth's dry land in its northern hemisphere while the land in the southern hemisphere was perpetually submerged.

In the fourteenth century, John Buridan had arrived at a similar solution, though in a quite different manner and without appeal to scriptural authority. Convinced that the spheres of earth and water were concentric with respect to the center of the world, Buridan assumed that the water did not completely surround the earth because some part of it flowed naturally downward and filled the bowels of the earth, while other parts of it mixed with air after evaporation.[75] The quantity of water was thus insufficient to cover the entire earth and so, inevitably, part of the earth was left exposed above the waters. Moreover, the exposed part would be rarefied and lightened by the sun's heat and the action of the air, whereas the submerged part would remain heavy and dense. Homogeneity for the earth was thus impossible, so that the earth's center of gravity must differ from its center of magnitude.[76] In this physical arrangement, only the earth's center of gravity was coincident with the center of the world.

Here Buridan anticipated a serious objection. Over long periods of time, geological processes—especially those wherein the waters flowing to the seas carry earthy matter down from the mountains—should wear down the mountains and elevations leaving the earth everywhere submerged

---

[74] Paul of Burgos's interpretation appears in his *Postillae Nicolai de Lyra super totam bibliam cum additionibus Pauli burgensis et replicis Matthiae Doringk* (Nuremberg, 1481) and is reported by Randles, *De la terre plate*, 29–30. For convenience, I have divided Burgos's diagram (as reproduced by Randles) into two parts.

THE GATHERING OF THE WATERS

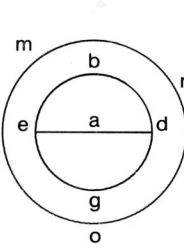
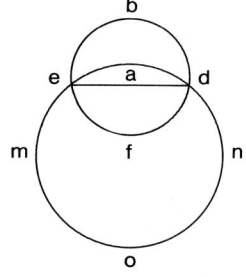

BEFORE	AFTER
Circle *ebdg* represents the earth and circle *mno* the surrounding sphere of water. Point *a* is the common center that coincides with the center of the world.	Arc *ebd* represents the portion of dry land elevated above the sphere of water *mno*; the center of the sphere of water, *f*, is now separated from the earth's center (and the center of the universe), *a*.

[75] Buridan, *Questions on De caelo* (Moody ed.), bk. 2, question 7 ("Whether the whole earth is habitable"), 159. For a translation, see Edward Grant, *A Source Book in Medieval Science*, 623 (most of the question is translated on pp. 621–624).

[76] In his *Questions on the Physics*, Bk. 4, question 5 ("Whether the earth is in water or in the concave surface of this water"), ed. cit. (above, n. 6), fol. 46r, col. 2, Albert of Saxony made the same distinction on much the same grounds.

below the waters.[77] Buridan explained how this potentially drastic consequence was avoided. The earthy matter continually deposited in the seas
by the waters flowing down from the mountains and elevations makes
the submerged portions of the earth heavier, which, in turn, causes the
earth's center of gravity to shift continually. With each such shift, the
earth moves rectilinearly until its center of gravity coincides with the center
of the world. These minute, but incessant, rectilinear shifts of the earth's
center of gravity will cause previously submerged parts of the earth to
rise above the surface of the seas and oceans.[78] Because this geologic
process is cyclic and continuous, part of the earth will always remain
elevated above the waters.

In other respects, however, Buridan's theory resembled that of Paul of
Burgos. Both were based on the assumption of two separate spheres, one
of earth, the other of water. The interrelations of the two spheres were
such that only a quarter of the earth's sphere was elevated above the
waters. The habitable portion of the earth's sphere lay wholly in the
northern hemisphere with the southern hemisphere completely submerged.

It was Albert of Saxony who, in his *Questions on the Physics*, may have
first proclaimed, and then perhaps abandoned, a new relationship between
earth and water. What lies at the geometric center of the world is not the
earth's center of gravity, but rather the center of "the whole aggregate of
earth and water, which make one whole heaviness, the center of gravity
of which is the center of the world."[79] From this relationship, however,
Albert did not infer or conclude that earth and water formed a single
sphere. Indeed from the discussion immediately following, Albert seems
to accept separate spheres for each.[80] At another time in his career, however,
Albert actually rejected this same opinion, as is evident in his *Questions
on De celo*, where he denied that "the center of gravity of the whole
aggregate of earth and water is the center [or middle] of the world."[81]

---

[77] Buridan (Moody ed.), 154–155; English translation, Grant, *Source Book,* 621–622.

[78] Buridan (Moody ed.), 159–160; English translation, Grant, ibid., 623. Part of this section
is translated into French by Randles, *De la terre plate,* 43.

[79] "Sexto dico quod conformiter intelligendum est de toto aggregato ex terra et aqua que
forte faciunt unam totalem gravitatem cuius medium gravitatis est medium mundi." Albert
of Saxony, *Questions on the Physics,* Bk. 4, question 5, fol. 46r, col. 1. The Latin text with
French translation is also cited by Duhem, *Le Système du monde,* Vol. 9, p. 213. On pp. 205–
219, Duhem discusses Albert of Saxony's concepts of the equilibrium of land and sea.

[80] This seems apparent from Albert's claim that water, rather than air, is the natural place
of the earth. If earth and water formed a single sphere, we would expect Albert to postulate
the air as the natural place of that single sphere. For Albert's discussion, see ibid., fol. 46r,
col. 2.

[81] In Bk. 2, question 23 ("Whether the earth is in the center [or middle] of the world") of
his *Questions on De celo,* fols. 116v, col. 2–117r, col. 1, Albert of Saxony insisted that water
cannot essentially affect the position of the earth. For if we imagine all the water removed,
the earth's center of gravity alone would coincide with the center of the world. Nor is the
earth's center of gravity pushed away from the center of the world by the weight of the
waters that rest on and cover the uninhabited side of the earth, as Buridan and Pierre d'Ailly
hold. In Albert's judgment, earth is *essentially* heavier than water, a fact made obvious by
the descent of a small piece of earth through a large body of water; or, to put it another
way, "water is essentially less heavy than earth." Since water cannot affect the behavior of

If Albert of Saxony appears ambivalent, Pierre d'Ailly (1350–1420) was not. Well acquainted with the opinions and theories of his Parisian predecessors, Buridan, Oresme, and Albert of Saxony, Pierre d'Ailly chose to support the interpretation that assigned a single center of gravity to the totality of earth and water. For virtually the same reasons as Buridan, d'Ailly distinguished between the earth's center of gravity and its center of magnitude.[82] Unlike Buridan, however, he denied that either the earth's center of gravity or its center of magnitude occupied the center of the universe.[83] What coincides with the center of the world is the center of gravity of *the aggregate* of earth and water (*centrum gravitatis aggregati ex aqua et terra*), a coincidence that arises from the natural tendency of every heavy body—and the composite of earth and water is a heavy body—to seek and remain at the center of the world.[84] Thus the earth does not lie at the center of the world *per se*, but only as part of a composite that also includes the waters within it and on its surface.[85] D'Ailly probably conceived the aggregate of earth and water as two separate, but interrelated, spheres, which taken together formed a heterogeneous mass with distinct centers of magnitude and gravity. D'Ailly thus retained the traditional belief that the southern hemisphere was covered by water and therefore uninhabited.

It was not until the Portuguese explorations of the southern hemisphere, especially the voyage along the coast of Brazil in 1501,[86] that Europeans learned of the wide distribution of land in that region. In a letter to Rudolf Agricola that was published in 1515 and a few times thereafter, Joachim Vadianus (1481–1551) of Switzerland took cognizance of the new knowledge to proclaim that not only did earth and water together form a single globe, but their relationship was such that over the entire surface of that

---

the essentially heavier earth, only the earth's center of gravity can occupy the center of the world, not that of the earth and water combined.

For Duhem's translation of this passage, see *Le Système du monde*, vol. 9, p. 215, where Duhem also notes Albert's drastic change of opinion from his *Questions on the Physics*. Which of the two opinions represents Albert's final judgment will only be known when the order of composition of his *Questions on De celo* and *Physics* is determined.

[82] Pierre d'Ailly, *Questions on the "Sphere" of Sacrobosco*, ed. cit., question 5, fol. 28r, col. 1. For Duhem's summary of d'Ailly's discussion, see *Le Système du monde*, 9: 231–235.

[83] Secunda conclusio: centrum gravitatis terre non est in medio firmamenti. Patet conclusio quia si terra ymaginetur dividi in duas partes eque graves tunc illa pars que est aquis cooperta una cum aqua circundante pellit aliam partem quousque centrum totius aggregati sit centrum mundi.

Tertia conclusio: est quod non idem centrum magnitudinis terre et firmamenti quia tunc terra esset omnino aquis cooperta . . .'' Pierre d'Ailly, ibid.

[84] ''Quarta conclusio est quod centrum gravitatis aggregati ex aqua et terra est in medio firmamenti. Patet quia tale aggregatum est corpus grave et non impeditum. Ergo movetur quousque centrum gravitatis eius sit centrum mundi.'' Ibid.

[85] D'Ailly distinguished four ways (ibid.) in which the earth could be in the center of the world: (1) by its center of magnitude; (2) by its center of gravity; (3) as part of an aggregate, the center of which is in the center of the world; or (4) because the earth is surrounded by the firmament. A number of lines below (ibid.), d'Ailly declares that ''the earth can be said to be in the middle [or center] of the firmament in the third way,'' a claim that also applies to water.

[86] See Randles, *De la terre plate*, 44–45.

globe earth is partly submerged and partly elevated.[87] Here was perhaps the first proper description of what would be called "the terraqueous globe" in the seventeenth century (see above, n. 69).[88] It was, as we saw, a conception that Copernicus would adopt. But Copernicus went further and insisted that within this globe the centers of gravity and magnitude were identical, a conclusion that committed him to a terraqueous sphere that was not only homogeneous in composition, but, as he would also declare, "is perfectly round, as the philosophers held."[89] For only if our sphere were "perfectly round" could the two centers be identical, although, in the heliocentric system, neither could coincide with the center of the world.[90]

With Christopher Clavius's *Commentary on the "Sphere" of Sacrobosco*, the terraqueous globe entered scholastic cosmology in the late sixteenth century[91] and was widely adopted in the seventeenth. Although there were important dissenters,[92] Aversa, Amicus, Cornaeus, Mastrius and Bel-

---

[87] Ibid., 45.

[88] Although he rejected it, John Buridan had already described a single terraqueous globe in the fourteenth century. The opinion he described assumed that earth and water had a common center which was identical with that of the center of the world. Moreover, those who hold this opinion say that "in any quarter of the earth there are many regions not covered by waters because of many protrusions of earth and elevations of mountains projecting above the waters. And [they] also say that many other parts of the earth are covered by waters because of their depressions, such as valleys between the aforementioned elevations. They say that is so in any quarter of the earth. The sign of this is that from one very large uncovered region, we cross a great and long sea and come to [yet] another very large uncovered region. It is probable that this would be the case as one went round the whole earth." My translation from Grant, *Source Book in Medieval Science,* 622. The passage occurs in Buridan's *Questions on De caelo,* Bk. 2, question 7 ("Whether the Whole Earth is Habitable"), Moody ed., 157. In reply, Buridan insists that only our quarter of the earth is habitable, with the rest submerged, as is evident from our inability to reach land beyond the Pillars of Hercules (the Straits of Gibralter).

[89] ". . . rotunditate absoluta, ut philosophi sentiunt." *Nicolai Copernici Thorunensis De revolutionibus orbium caelestium libri sex,* edited by F. and C. Zeller (Munich, 1949), Bk. 1, ch. 3, p. 12, line 23. The translation is by Rosen, *Nicholas Copernicus On the Revolutions,* 10, line 25.

[90] Goldstein observes ("The Renaissance Concept of the Earth," *Terrae Incognitae,* 4, 41, n. 41) that for Copernicus, the earth's perfect rotundity and integrity were necessary preconditions for its revolution around the sun. In Copernicus' mind, the daily rotation was also probably dependent on the same conditions. For more on the significance of Copernicus's departure from Aristotle's two sphere system of earth and water, see Goldstein, ibid., 47–48.

[91] For Clavius's lengthy discussion, which occurs under the title "An ex terra et aqua unus fiat globus, hoc est, an horum elementorum convexae superficies idem habeant centrum," see *In Sphaeram Iohannis de Sacro Bosco Commentarius* (Lyon, 1593), 133–151.

[92] Aversa (*Philosophia,* 225, col. 2) would cite the Conimbricenses (*De caelo* commentary, Bk. 2, ch. 14, question 3; in the Lyon edition of 1598 [see above, n. 52], see Bk. 2, ch. 14, article 2 ["Centrum mundi, centrum gravitatis et magnitudinis terrae esse unum et idem"], pp. 382–384) and Ruvio (*De Caelo* commentary, Bk. 3, ch. 8, question 2) as among those who held that the earth was distinct from the waters that surrounded it. According to Aversa, they argued that the heaviest natural body, or the earth, was assumed to occupy the lowest, or most central, place in the universe, which would be impossible unless the earth's center of gravity coincided with the center of the universe. Only water could prevent the earth's center of gravity from coinciding with the center of the universe. This could be achieved in

lutus would not be among them.[93] They assumed that earth and water formed a single globe, which Aversa and Mastrius and Bellutus called "the terrestrial globe" (*terrestris globus*).[94] Numerous arguments were proposed in favor of a single globe of earth and water. A popular astronomical argument invoked the round shadow cast upon the moon during a lunar eclipse. Because of its surface irregularities, the earth alone was thought incapable of casting such a shadow. Only the combination of earth and water (oceans and seas) could constitute a sufficiently round globe to cast a round shadow on the moon.[95] Another argument relied on the experiences of sailors.[96] When ships are at sea away from land, sailors see nothing but sky and water. As they approach land, however, they see the summits of mountains first, then the middles, and finally the bottoms of those same mountains. Such a gradual appearance of mountains from top to bottom was for Amicus evidence that earth and water have a continuous convex surface.[97] For if not, and water were elevated above the earth, sailors would see things on shore all at once rather than gradually.[98] Moreover, if earth and water constituted a single globe, they must have the same

---

one of two possible ways: by nature or by force. The former would be impossible because water naturally tends to rise above the earth; the latter would be ineffective because of the Aristotelian principle that anything caused by violence must be temporary. For Aversa's specific arguments against the Conimbricenses and Ruvio, see Aversa, ibid., 227, col. 2–228, col. 1.

[93] See Aversa, *Philosophia*, 227, col. 2–228, col. 1; Amicus, *De coelo*, tract 8, doubt (*dubitatio*) II ("An centrum magnitudinis terrae sit idem cum centro gravitatis"), 582, col. 2–585, col. 1, 598, col. 1; Cornaeus, *Curriculum philosophiae peripateticae*, 518–520; and Mastrius and Bellutus, *De coelo*, 560, col. 1–561, col. 2.

[94] Aversa used the expression frequently, as, for example, on 224, cols. 1, 2; 226, col. 1; 227, col. 1; 231, col. 1; and 232, cols. 1, 2 of his *Philosophia*. For instances of its usage by Mastrius and Bellutus, see *De coelo*, 562, col. 1, par. 113 and 564, col. 1, par. 122. It is thus necessary to revise the claim by Randles (*De la terre plate*, 63) that the term "terraqueous globe" was not replaced by the term "terrestrial globe" until the eighteenth century.

[95] See Amicus, *De coelo*, 583, col. 1 and Cornaeus, *Curriculum philosophiae peripateticae*, 518. This argument was undoubtedly derived from Clavius, *In Sphaeram Iohannis de Sacro Bosco Commentarius*, (Lyon, 1593), 140.

[96] Amicus, *De coelo*, 583, col. 1.

[97] In his lengthy discussion on "whether there is one globe of earth and water," Clavius demonstrated that "earth and water have one and the same convex surface and, consequently, the same center" (see *In Sphaeram Iohannis de Sacro Bosco Commentarius*, 139). Although Mastrius and Bellutus (*De coelo*, 559, col. 2, par. 100 and 560, col. 1, par. 102) held that earth and water make one globe and are almost equal in "height" (*altitudo*), they concluded that, because of its mountains, the earth is overall slightly higher than water, a position that was also held by Duns Scotus.

[98] As with other arguments, Amicus apparently derived this one from Clavius since he mentions (ibid.) that Francesco Patrizi (in the sixth book of his *Pancosmia*) had attempted to refute it in the form presented by Clavius (for Clavius's version, see his *In Sphaeram Iohannis de Sacro Bosco Commentarius*, 140–141. On 586, col. 1–587, col. 2 (in a section titled "whether the earth is higher than the sea, or more depressed") Amicus proposed additional arguments that the seas and oceans were not elevated above the surface of the earth. Aversa (*Philosophia*, 226, col. 2) argued that if the earth and water were separate spheres and the earth's center coincided with the center of the universe, the water on one side of the earth would be elevated above the earth's surface. We should therefore expect this water to flow down by its very nature unless prevented by a perpetual miracle. Under these circumstances, rivers would not flow down to the sea but ascend to it.

center, which Amicus demonstrated as follows:[99] if a part of water and a part of earth were dropped in air over the same path, experience indicates that they fall toward the same center. Indeed if two such heavy bodies that were dropped from the same place descended over different paths, they would tend toward different centers. But this is contrary to experience. In fact, if water and earth formed two different globes or spheres, two heavy bodies, one earth and the other water, could not tend toward the same center because one globe would intersect the other and they would not possess a single center, just as two circles that mutually intersect cannot possess the same center.[100]

From these and similar arguments, most scholastic natural philosophers of the seventeenth century assumed that earth and water formed a single sphere. But just as those who earlier had assumed that the earth alone was the sphere at the center of the world, the supporters of the terraqueous sphere had to solve the problem of the three centers, that is, they had to determine whether or not the center of the universe and the centers of gravity and magnitude of the terraqueous sphere were identical. Fundamental to this determination was the conclusion that the terraqueous globe had a single, convex surface composed of water and earth. Thus the compound globe was not conceived as a mass of water opposed to a mass of earth, but the two elements were thought of as commingled throughout the extent of the globe.[101] Water was assumed to intrude into numerous cavities within the earth, while earth, in the form of islands and peninsulas, was distributed throughout the oceans and seas. Because of the earth's mountains and prominences, the unified terraqueous globe was not considered a perfect, geometrical sphere.[102] Clavius had nevertheless insisted that the inequality or difformity was negligible when compared to the entire globe. The description of the globe as round and spherical was therefore justified,[103] from which Clavius concluded that the center of the universe and the centers of gravity and magnitude of the terraqueous sphere are one and the same. Aware of this, Aversa nevertheless recounted the arguments of those who denied such a coincidence of centers.[104] A globe composed of earth and water, where the water is admittedly lighter than earth, could no more have the same centers of gravity and magnitude than a ball made partly of lead and partly of wood. Indeed the mountains

---

[99] Amicus, *De coelo*, 582, col. 2–583, col. 1.

[100] Here Amicus seems to assume that the two separate spheres would intersect, presumably because parts of the earth rise above the seas and oceans.

[101] Aversa, *Philosophia*, 228, col. 2–229, col. 1; Amicus, *De coelo*, 588, col. 1; and Cornaeus, *Curriculum philosophiae peripateticae*, 519. For Clavius's discussion, see *In Sphaeram Iohannis de Sacro Bosco Commentarius*, 142–143.

[102] Amicus, *De coelo*, 587, col. 1 and Cornaeus, *Curriculum philosophiae peripateticae*, 517–518. Since Mastrius and Bellutus believed the earth was slightly higher than the water that with it made up the terraqueous sphere, it is obvious that they did not consider the composite sphere a mathematical sphere (see above, n. 97).

[103] *In Sphaeram Iohannis de Sacro Bosco Commentarius*, 141–142.

[104] Aversa, *Philosophia*, 229, col. 1.

on the exposed parts of the earth would appear to make that part of the globe heavier than the rest of it. The same dilemma remains if, to the contrary, one accepts Pierre D'Ailly's argument that the watery parts of the globe are heavier than the earthy parts because significant masses of the latter are exposed to the drying action of the sun and are therefore lighter.[105]

Aversa conceded that the terraqueous globe is unequal in mass and weight so that the same point could not function as its center of magnitude and gravity. He concluded that only the center of gravity coincided with the center of the universe whereas the center of magnitude was slightly off center. But like Clavius, Aversa thought the difference between the centers of magnitude and gravity so slight as to be negligible and concluded, as did Clavius, that all three centers are identical.[106] As if to reinforce his conclusion, Aversa emphasized the balance between water and earth in the terraqueous sphere: the heaviness caused by an abundance of waters on one side of the earth was counterbalanced on the other side by the mountains that rise high above it.[107]

The terraqueous, or terrestrial, globe of earth and water was assumed to lie naturally at the center of the universe. As Clavius explained, Sacrobosco had demonstrated in the thirteenth century that the earth is situated in the center of the universe. However, when Sacrobosco speaks of "the earth," we must understand earth and water.[108] Physically, the compound terraqueous globe lies at the center of the universe because both earth and water, as heavy bodies, move down naturally to seek the center of

---

[105] Aversa's reference to D'Ailly is to the latter's *Questions on the "Sphere" of Sacrobosco,* ch. 1, question 5, article 3. D'Ailly's opinion bears a striking resemblance to John Buridan's discussion in the latter's *Questions on De caelo,* Bk. 2, question 7 ("Whether the whole earth is habitable"), 159 of Moody's edition. For a translation of the passage, see Grant, *Source Book in Medieval Science,* 623.

[106] Aversa, *Philosophia,* 230, cols. 1–2. Although he often used the term "earth" (*terra*) for the terraqueous globe, Cornaeus (*Curriculum philosophiae peripateticae,* 519–20) also assumed that the centers of the earth's gravity and magnitude were coincident with the center of the universe. Or, as he expressed it, "it is sufficiently certain that the centers of gravity and magnitude coincide physically in the earth and do not differ noticeably" (519), a claim based on our "experience that the globe of earth is uniformly mixed with water." That the center of gravity (and, therefore, also the center of magnitude) also coincides with the center of the universe is demonstrated (520) from the fact that "the earth, as the heaviest of all bodies, tends toward the center of the whole universe, which it does as the point of its gravity is carried perpendicularly and occupies the center of the world by virtue of its heaviness. But it could not occupy the center of the world with its gravity unless its center of gravity were fixed in the center of the world." For reference to the relevant passages in Clavius, see above, n. 103. Mastrius and Bellutus also argued (*De coelo,* 561, col. 1, par. 108) that the three centers were physically, though not mathematically, coincident. For if the centers were mathematically identical, the slightest variation in any part of the earth—for example, the fall of a small stone on to the earth's surface—would cause the earth's center of gravity to shift slightly from the mathematical center. Such a slight shift would not, however, noticeably affect the physical—that is, approximate—coincidence of the three centers.

[107] Aversa, ibid., 230, col. 2–231, col. 1. Amicus (*De coelo,* 588, col. 1) and Cornaeus (see n. 106) also emphasized a balance between water and earth.

[108] Clavius, *In Sphaeram Ioannis de Sacro Bosco Commentarius,* 151.

the universe.[109] That they seek the center of the universe was evident to Aristotle, who, as Amicus observes approvingly, argued that the downward motions of heavy bodies were always toward the same center at equal, not parallel, angles, that is, at right angles to a tangent.[110]

The earth's centrality was also supported with appeals to astronomical evidence. Most of these arguments were drawn from Clavius, who declared that if the earth were on the axis of the world and in the plane of the celestial equator, it would lie exactly at the center of the world. But if the earth were not at the center of the universe, as the Copernicans argued, then three possibilities were open: (1) the earth could lie in the plane of the celestial equator but outside the axis of the world; (2) it could lie on the axis of the world but not in the plane of the celestial equator; (3) or the earth might be located neither on the axis of the world nor in the plane of the equator. For all these possibilities, Clavius derived consequences that were contrary to experience.[111] Scholastic authors seized upon some of these as evidence for the earth's centrality. Most frequently mentioned was the fact that from any place on earth six signs of the zodiac were always visible with the other six invisible.[112] For if the earth were not in the center of the world, but nearer to one side of the spherical heaven than to another, anyone on that side of the earth's surface nearer to the heaven would observe less than half of the heaven; and if he were on the opposite side of the terrestrial surface, he would see more than half of the heaven. Under these circumstances, no great circle could divide the heaven into two equal parts, which is contrary to experience.[113] Another frequently mentioned consequence showed that if the earth lay outside the center of the world, the same stars ought to appear larger when nearer the earth and smaller when farther away.[114] According to Aversa, the same consequence would follow if the earth moved in the great orb assigned

---

[109] Mastrius and Bellutus, *De coelo*, 559, col. 1, par. 97; Amicus, *De coelo*, 581, cols. 1–2; Cornaeus, *Curriculum philosophiae peripateticae*, 534.

[110] Amicus, *De coelo*, 581, col. 1, who cites Aristotle, *De caelo*, Bk. 2, Text 102 (2.14.296b.15–21).

[111] Clavius, *In Sphaeram Ioannis de Sacro Bosco Commentarius*, 154–160. Citing Clavius, as he frequently did, Amicus (*De coelo*, 581, col. 1) repeats the same three alternatives.

[112] Aversa, *Philosophia*, 5, col. 2; Amicus, *De coelo*, 581, col. 1; Mastrius and Bellutus, *De coelo*, 482, col. 1, par. 17; Clavius, *In Sphaeram Ioannis de Sacro Bosco Commentarius*, 152–153, 154–155. Sacrobosco's *Sphere* was the ultimate source of this argument (see Thorndike, *The "Sphere" of Sacrobosco and Its Commentators*, 84 [Latin], 122 [English]). It also appears among the 195 arguments, or "assertions", against the Copernican system assembled by the Jesuit Giorgio Polacco in his *AntiCopernicus Catholicus, seu De terrae statione, et de solis motu, contra systema Copernicanum, Catholicae assertiones* (Venice, 1644), 104 (Assertio CXC).

[113] See Amicus, ibid., who derived it from Clavius (ibid., p. 153). For much the same argument, see also, Aversa, ibid., 5, col. 2, who specifically mentions that this untenable consequence would follow if the earth were really in the great orb described by Copernicus ("si esset in orbe illo magno iuxta descriptionem Copernici").

[114] Mastrius and Bellutus, *De coelo*, 482, col. 1, par. 16. Clavius makes the same point (ibid., 151).

to it by Copernicus; the same star ought to appear larger on one side of the earth's orbit and smaller on the opposite side.[115]

If the earth were eccentric, Aversa insisted[116] that other absurd and contrary to fact conditions could be derived "from the equinoxes and solstices, from the increase and decrease of the days and nights, from the relationship of the horizon between heaven and earth, from the existence of the solar shadow, from a lunar eclipse,[117] and from many other such observations, which require further astronomical investigation." In Aversa's judgment, it was because of these difficulties "that Copernicus made the height of the starry heaven enormously far away, no doubt so that this whole great orb [of the earth's annual motion] would be as a point with respect to it [i.e. the starry heaven] thus producing no sensible diversity." But "this clearly exceeds all faith and is contrived beyond reason and yet does not properly suffice to save all the appearances."

---

[115] Aversa, *Philosophia*, 5, col. 2–6, col. 1.

[116] Ibid., 6, col. 1.

[117] With bk. 2 of Averroes's *Commentary on De caelo* as his explicit source, Amicus argued (*De coelo*, 581, col. 1) that if the earth is not at the center of the world, lunar eclipses could not occur when the Sun and moon are diametrically opposed. With the earth outside the center of the world, the earth could not prevent the Sun's illumination from reaching the moon when Sun and moon are at opposite points of the Zodiac.

IV. THE EARTH'S IMMOBILITY

Although we have seen that a few Aristotelians were prepared to allow certain motions to the earth, it is obvious that most were committed to its immobility at the center of the world. With the condemnation of the Copernican system in 1616, some Catholic Aristotelians, especially scholastic theologians concerned with cosmology and physics, chose to include a brief discussion of that system with subsequent refutations of its major claims. The authors on whom this study concentrates did so and probably represented a majority among scholastics who wrote on cosmology and physics between 1616 and 1665.[118]

Already in the sixteenth century the various Greek and Latin reports of the different possible terrestrial motions—including Archimedes's report of Aristarchus's heliocentric system[119]—made discussion of that theme difficult to avoid in comprehensive physical and cosmological treatises and commentaries. With the condemnation of Copernicus's *De revolutionibus* in 1616, discussions of the possibility of one or more actual terrestrial motions became *ipso facto* controversial and significant for much of the remainder of the seventeenth century. Although not all scholastics would mention Copernicus in their discussions of the earth's possible motions,

---

[118] Among those who published after 1616 and discussed the earth and one or more of its motions, but make no mention of Copernicus, we may include: Roderigo de Arriaga, S. J. (1592–1667), *Cursus philosophicus* (Antwerp, 1632) (for Arriaga's consideration of the earth's center of gravity and its motions, see 587–588) and Petrus Hurtadus de Mendoza, S. J. (1578–1651), *Universa philosophia, nova editio* (Lyon, 1624; for Hurtado's discussions of the earth's centers and motions, see 383–386). Sigismund Serbellonus, a Professor of Theology in Milan, mentions "the damned opinion of Copernicus" and describes the latter's system with the sun at rest in the center of the world and the earth placed in the epicycle of the moon in the third heaven. But he immediately replies that "the common opinion remains firm concerning the immobility of the earth in the center of the world [and] the revolution of the planets and the firmament, or *primum mobile*" (see *Philosophia Ticinensis R. P. D. Sigismundi Serbelloni Mediolanensis ex clericis regularibus Barnabitis Congreg. S. Pauli*, vol. 2 [Milan, 1663], 39, col. 1). Despite mention of Copernicus and Tycho Brahe (ibid., 28), Serbellonus says nothing more about the earth's possible motions. Although we shall concentrate on the opinions of Amicus, Aversa, Cornaeus, Mastrius, Bellutus, and Riccioli, who specifically mentioned Copernicus, authors like Arriaga and Hurtado, as well as relevant authors of the late sixteenth century, will also be cited.

[119] For a translation of Archimedes's description in the *Sandreckoner*, or *Arenarius*, see Sir Thomas Heath, *Aristarchus of Samos, The Ancient Copernicus* (Oxford: at the Clarendon Press, 1913), 302. Heath observes that Plutarch (*Concerning the Face Which Appears in the Orb of the Moon* [*De facie quae in orbe lunae apparet*], 922F–923A) added the important detail that Aristarchus also assumed a daily axial rotation of the earth. Heath's translation is reprinted in Morris R. Cohen and I. E. Drabkin, *A Source Book in Greek Science* (Cambridge, Mass.: Harvard University Press, 1958; first edition, McGraw-Hill Book Co., 1948), 108–109.

they were heirs to a long tradition of arguments about the earth's motions that traced back to the thirteenth and fourteenth centuries and also embraced the more recent sixteenth-century additions to the stock of authors and arguments. As Amicus expressed it, "In our age, Copernicus has raised this opinion, which had been buried with the dead, in a work *De revolutionibus orbium caelestium*, where he says that the heavens are immobile and the earth is moved."[120] Copernicus raised this old, but not quite dead, issue by the assumption of an "immobile firmament, with the sun fixed in the center of the universe, and the earth, which is moved by a triple motion by [means of] which he attempts to save all the appearances, in the third heaven."[121] The three terrestrial movements to which Amicus alludes were (1) a daily axial rotation; (2) an annual west to east motion around the sun; and (3) what may best be described as "other motions," which, depending on the author, could be a rectilinear movement, a trepidational, or axial, motion, and even earthquakes.[122]

Of the three motions, the daily rotation attracted the most attention. With an occasional exception, such as Riccioli, most scholastics considered the daily and annual motions together, although they concentrated on the daily rotation.[123] In this study, we shall summarize the major scholastic arguments for and against the earth's immobility, emphasizing primarily the physical arguments and, to a lesser extent, the metaphysical and theological responses.[124]

A. PHYSICAL ARGUMENTS BASED ON THE COMMON MOTION

1. The Common Motion

None of the physical consequences derived from acceptance of the earth's axial rotation was more pervasive and perplexing than that of the common motion, which, in order to save a wide range of physical phenomena, assumed that all bodies on and above the earth's surface shared in the earth's rotational motion. Ptolemy had already used the concept of common motion to subvert belief in the earth's daily rotation,[125] while,

---

[120] Amicus, *De coelo*, 289, col. 1.

[121] Ibid.

[122] While Cornaeus (*Curriculum philosophiae peripateticae*, 530) assumed that the third motion was that which Copernicus adopted, namely a libration of the earth's axis from south to north for six months and from north to south for another six months (see Rosen [tr.], *Nicholas Copernicus On the Revolutions*, Bk. 1, ch. 11, 23–25 and 360 for Rosen's quotation from Kepler's *New Astronomy* on the function of that motion), Aversa identified the third motion with an actual downward rectilinear movement (*Philosophia*, 231, col. 1), which he derived from Seneca's *Natural Questions*, Bk. 7, ch. 14.

[123] Even Riccioli occasionally formulated arguments against both motions in the same paragraph or section.

[124] Except for those mentioned earlier, the astronomical arguments will be excluded.

[125] *Almagest*, Bk. 1, ch. 7, translated by R. Catesby Taliaferro, *Great Books of the Western World* (Chicago: Encyclopaedia Britannica, 1952), vol. 16, p. 12. Reprinted in Grant, *A Source Book in Medieval Science*, 495–496.

in the fourteenth century, Nicole Oresme, in the course of a series of hypothetical arguments in favor of the earth's rotation, defended it as plausible.[126] It was, however, Copernicus's version of the common motion argument that became the focal point of scholastic reaction in the seventeenth century. In *De revolutionibus*, Bk. 1, ch. 8, Copernicus declared "that the motion of falling and rising bodies in the framework of the universe is twofold, being in every case a compound of straight and circular."[127] To justify terrestrial rotation, Copernicus abandoned the Aristotelian idea that rectilinear motion was natural for the four elements. Indeed natural motion for earth and fire was circular as long as each of these elements was united to its whole. Only "when they are separated from their whole and forsake its unity,"[128] do they move rectilinearly. Although detached from the earth, watery and earthy things in the air, and the air itself, share in the earth's rotational motion.[129]

The scholastic arguments against the Copernicans have a familiar Ptolemaic ring. There were those, like Raphael Aversa, who argued against the earth's rotation as if no one had ever proposed the common motion argument. The earth's immobility, Aversa insisted,[130] could be demonstrated from a variety of experiences. If the earth really turned daily from west to east, the clouds would appear "to be carried constantly from east to west and in no way to remain over the same place of the earth."[131] When anyone projects a stone upward with great force, it ought to fall to the earth considerably to the west "because the motion of the earth has, in the interim, continued from west to east." But "unless it was not projected in a straight line or was moved somewhat by the agitation of the air, the stone falls back to the same place." Finally, if the earth rotates swiftly from west to east, we should feel a strong wind from east to west. But no such effect is perceived.[132] For all these reasons, then, "it is surely not the earth that is revolved constantly with a daily motion."

Although such arguments were frequently repeated, many scholastics were aware that the Copernicans had attempted to meet them by assigning

---

[126] *Nicole Oresme: Le Livre du ciel et du monde*, Bk. 2, ch. 25, 525–527 of the edition by A. D. Menut and A. J. Denomy (1968); the relevant section is reprinted in Grant, *A Source Book in Medieval Science*, 505–506.

[127] Rosen (tr.), *Nicholas Copernicus On the Revolutions*, 16.

[128] Ibid., 17.

[129] Ibid., 16. Galileo's similar arguments about the common motion (see the Second Day of his *Dialogue Concerning the Two Chief World Systems* [*I due massimi sistemi del mondo Tolemaico e Copernicano* [Florence, 1632]) were read by some scholastics, among whom we may include Riccioli, Mastrius, and Bellutus.

[130] For the arguments cited from Aversa, see his *Philosophia*, 143, col. 2.

[131] Amicus (*De coelo*, 289, col. 2) mentions the same argument but omits the Copernican rebuttal. Galileo also mentioned it in the *Dialogue Concerning the Two Chief World Systems*, trans. Stillman Drake (Berkeley and Los Angeles: University of California Press, 1962), The Second Day, 131–132.

[132] Riccioli includes this argument as the twenty-eighth in favor of the earth's immobility (for the reference and further discussion, see below, n. 134). Galileo also included it in his *Dialogue Concerning the Two Chief Systems*, The Second Day, 132.

the earth's rotational motion to all things in the air above the earth's surface. In his summary of arguments in defense of the earth's immobility, Riccioli was content to cite these and similar arguments, but in each case concluded with the Copernican response based on the common motion.[133] For example, he described the claim that if the earth rotates from west to east, we should have greater difficulty moving toward the west because of the air's resistance as the earth sweeps past. Riccioli then adds that the Copernicans deny that such a resistance would develop because "there is a common motion toward the east for bodies similar to the earth, just as with the air near the earth."[134] In these particular instances, though not in many others, Riccioli apparently chose not to resolve the argument in favor of the earth's immobility.

The Copernican arguments were nowhere better represented than in Galileo's *Dialogue Concerning the Two Chief World Systems.* After initially defending the earth's immobility by appeal to various experiences but without mentioning the typical common motion arguments of the Copernicans, Mastrius and Bellutus[135] invoked certain of Galileo's arguments which depended on the assumption that all things shared a common, circular motion. These arguments had been introduced by Galileo in order to refute the so-called absurdities alleged to follow from the daily rotation by opponents of the Copernican system.[136]

2. Ships and the Common Motion

Of particular interest to Mastrius and Bellutus was Galileo's discussion of the motions of various animate and inanimate things located within

---

[133] See the twelfth to fifteenth arguments in *Almagestum novum,* pars posterior, 474, col. 1.

[134] See the thirteenth argument. In his twenty-eighth argument (475, col. 1) in favor of the earth's immobility, Riccioli does much the same thing. There he not only mentions a perpetual wind toward the west, which was commonly cited, but adds that we should also perceive "sounds and hissing from the air striking against trees, mountains, towers, etc." Since we do not perceive such things, the earth must rest. But, in the conclusion of his argument, Riccioli seems to defer to the Copernicans by noting that they deny the occurrence of such effects by their insistence that the air near the earth, which is filled with exhalations and vapors, would move with the common motion of the earth.

[135] Mastrius and Bellutus, *De coelo,* 562, col. 1, par. 113.

[136] Ibid., 562, cols. 1–2, par. 114. According to Mastrius and Bellutus, Galileo denied these absurdities "because not only is the earth moved innately (*ab intrinseco*) around the center with a circular motion, but also all bodies, whether animate or inanimate, whether united to the earth or separate, that exist in this elementary universe have this motion perpetually [and] innately so that they move simultaneously with the earth around the center of this elementary world. And because this motion is common to all, it is not perceptible to us except in relation to the fixed stars to which it does not apply." The arguments from Galileo are drawn from The Second Day of the latter's *Dialogue Concerning the Two Chief World Systems* published in 1632. My references to that work will be to Stillman Drake's translation (Berkeley and Los Angeles: University of California Press, 1962). For statements of the common motion, see 116, 163.

the confines of a cabin below the decks of a ship. Galileo insisted that in such a cabin, the observed motions of flies, men, fishes, and water dripping from one vessel to another would be the same whether the ship was at rest or in motion, provided only that the ship's motion was uniform and without fluctuations.[137] From this example, Mastrius and Bellutus explained the consequences derived by Galileo in defense of the earth's rotation and against all the absurdities that had been used against it.[138] A stone projected upward in the cabin would fall at the projector's feet because both projector and stone are moved with the ship. The stone, however, does not fall with a perpendicular motion but follows the path of a slanting line (*linea transversalis*) derived from the perpendicular motion of the stone and the horizontal motion of the ship.[139] In the same manner, an arrow projected upward would move with the motion of the rotating earth and fall at the foot of the projector. Indeed we are told that Galileo declared that "he himself had experienced many times that a stone projected from the top of a mast always fell to the foot of the mast, never into the sea, whether the ship was at rest or was moved very quickly."[140] The remainder of the discussion concerns the behavior of the different entities in the enclosed cabin. The common motion of the ship guarantees that their movements within the cabin will be the same as when the ship is at rest. At the conclusion of their presentation of Galileo's defense of the daily rotation, Mastrius and Bellutus attribute to Galileo the opinion that "this motion of the earth appears perplexing (*implicantem*) to us because we first conceive that the earth is immobile and from this assumption, which ought to be a conclusion that is proved, we proceed to infer many absurd things from which we then deduce the immobility of the earth by proving the same thing from the same thing from first to last."

---

[137] For Galileo's argument, see *Dialogue*, 186–187. Some 250 years earlier, Oresme made substantially the same argument (see Oresme, *Le Livre du ciel et du monde*, Bk. 2, ch. 25, 525 of the Menut translation; also Grant, *Source Book in Medieval Science*, 505).

[138] Although Galileo did indeed say that his cabin experiment would nullify all experiments that had previously been brought against the earth's rotation, Mastrius and Bellutus included consequences that Galileo did not specifically describe but which are compatible with his experiment.

[139] Although Galileo did not describe the path of a descending body in the cabin of a moving ship as "slanting," he earlier declared that the path of a falling stone toward a rotating earth would be slanting (see *Dialogue*, 173). Tycho Brahe had argued that an object hurled upward inside a ship would not fall to the same place regardless of the ship's rest or motion. The greater the ship's velocity, the greater the distance that would separate the places where the object would drop when the ship was at rest as compared to when it was in motion (see Ronald J. Overmann, "Theories of Gravity in the Seventeenth Century" [Ph.D. dissertation, Indiana University, 1974], p. 14, where the passage from Tycho's *Epistolarum astronomicarum liber primus* is translated from *Tychonis Brahe Dani Opera omnia*, vol. 6, ed. J. L. E. Dreyer [1919], 220, lines 16–21.

[140] Salviati, Galileo's spokesman, says this in the *Dialogue Concerning the Two Chief World Systems*, The Second Day, 144 (Drake trans.).

Although it was the point of departure, and even the basis for their entire discussion, Mastrius and Bellutus did not include any direct refutation of Galileo's argument about the various animate and inanimate objects in the enclosed cabin of the moving ship. Instead they focused attention on Galileo's fundamental Copernican claim that sublunar bodies possess an innate tendency to move around the center of the earth.[141] By this inherent property, the earth rotates around its own center in the same manner as do watery, airy, and fiery bodies. From the obvious fact that each type of sublunar body differed in species and genus from all the others, the falsity of such a claim seemed obvious to Mastrius and Bellutus. Such specific and generic differences precluded any common, innate tendency for rotatory motion.[142] Moreover, if circular motion were innate to the earth and all its parts, they should all possess an innate and determinate velocity—that is a unique speed. A single illustration with a stone should reveal the obvious falsity of such a consequence. "For if a stone were under the equinoctial [or equator], it ought to be moved most quickly, and as it is further removed from the equinoctial [and] toward the poles, it should be moved slower, so that if it were under the poles only it would be turned circularly in itself [i.e., it would rotate in position] and thus, in a space of 24 hours, it would complete the daily motion simultaneously with the whole earth." Thus sublunar bodies would travel with different speeds at different times, a condition that could not be produced by innate tendencies. Nor indeed could external causes produce a single, common rotatory velocity. Four such causes—angels, the place where a body moves, the nature of the whole, and air—were suggested and rejected by Mastrius and Bellutus. Leaving aside angels,[143] our authors reject the place of the stone because at best a place could only cause a single velocity. For if a stone were at the equator it would be moved with its greatest velocity, but as it is removed farther from the circle of the equator, its speed diminishes because it will traverse smaller circles in 24 hours. But although a place, such as the equatorial circle, might be the cause of the stone's quickest velocity, it could not be the cause of the stone's slower velocities at other parallels of latitude.[144]

---

[141] Ibid., 562, col. 2-563, col. 1.

[142] But apparently it did not preclude a common tendency for rectilinear motion, which Mastrius and Bellutus, and virtually every Aristotelian, would have thought true.

[143] The angel argument seems to require that angels be distributed throughout the whole orb (whether sublunar, or simply terrestrial is left ambiguous) so that "they could provide for moving a stone with a velocity proportioned to the whole motion as soon (statim ac) as it would be projected upward." Angels would have to be everywhere on the earth's surface ever ready to impose the proper rotational speed on every body hurled aloft. In this way, every stone, whether wholly or partly in the air (a stone needs no angelic assistance when on the earth's surface since the latter carries it through the daily rotation), would, presumably, be "programmed" by its attendant angel to complete its rotation in precisely 24 hours.

[144] Mastrius and Bellutus do not consider the possibility that each parallel of latitude might cause a specific degree of velocity and that the velocities diminish as the latitude increases.

The "nature of the whole" as an external cause of the daily rotational motion of all sublunar bodies detached from the earth presupposes that a rotating earth somehow communicates its property of rotation to every one of its parts. The properties that such a nature would require led Mastrius and Bellutus to reject it. Such a nature would have to be voluntary because it must somehow adjust the velocity of every stone hurled aloft in order to guarantee its rotation in 24 hours. After all, a stone could be moved with other motions either east or west. In the former case, the stone would complete its daily circulation in less than 24 hours; in the latter case, more than 24 hours. Hence the "nature of the whole," which presumably resides throughout the sublunar region, would have to regulate the velocities of all bodies moving above the earth's surface to insure for each precise 24-hour circulation. Finally, air as the external cause of the earth's daily rotation is unacceptable because "air cannot move a great mass with circular motion." Indeed if air pushed the earth in a daily rotation, the rotational speed of the air would be less than that of the earth because "the body that is pushing [namely, the air] and the body that is pushed [namely, the earth] are not moved equally." From the force of these arguments, Mastrius and Bellutus concluded that "this circular motion is assumed falsely and gratuitously to be innate in sublunar things."

Where Mastrius and Bellutus invoked Galileo's argument about animate and inanimate objects in the enclosed cabin of a moving ship merely as a point of departure, Giovanni Baptista Riccioli not only agreed with it but cited it in favor of the earth's immobility. He found occasion to introduce it following a discussion of the following proposition:

If the earth were moved with a daily rotation, or even an annual translation, the clouds hanging in the air, the smoke that rises, and the birds that are suspended [in the air], or flying toward the east, would always seem to be carried toward the west. But this is contrary to manifest experience. Therefore the earth is not moved with a daily rotation and much less with an annual translation.[145]

After a brief consideration of this claim, Riccioli invokes Galileo's argument about the objects in the enclosed cabin and does so with apparent approval.[146] For if the motions of the animate and inanimate objects in the enclosed cabin are precisely the same whether the ship moves or rests, a consequence that follows from the fact that the rest and motion of the ship are common to all, one may not infer the rest or motion of the ship from the motion of the objects in the cabin. In citing Galileo's argument, Riccioli may have subverted his own proposition about the birds in the air. By his apparent approval of Galileo's argument, Riccioli had, in effect, conceded that even if the birds do not appear to move toward the west

---

[145] Riccioli, *Almagestum novum*, pars posterior, 423, col. 2.
[146] Ibid., 424, col. 1.

when actually flying eastward, we cannot properly infer from this that the earth is at rest. The birds, after all, may appear to move eastward rather than westward either because the earth is really at rest, or because the birds share the eastward motion of a rotating earth, as Galileo and the Copernicans believed. And yet, in the proposition cited above, Riccioli did indeed infer the immobility of the earth from the eastward flight of the birds. Here again, we have reason to ponder Riccioli's motives. Why, after enunciating an argument in favor of the earth's immobility, did he introduce a Galilean argument that demonstrated the inconclusiveness of that same argument?

The common motion, or earth's axial rotation, was frequently denied on the basis of powerful appeals to sense experience. Riccioli emphasized that heavy bodies "descend naturally by a straight line perpendicular to the earth" and that if projected upward, "they would return over the same path to the same place." So obvious was this experience that it "could not be shown to be false by any more certain sensations, nor by any necessary a priori arguments, nor from things revealed by God."[147] Now there are only two possibilities: either heavy bodies descend in a path that is a straight line or they descend by means of a non-rectilinear line that only appears rectilinear. Those who argue against the senses and experience insist that the senses are false and misleading. Indeed they hold that such a judgment must not be made on the basis of the senses. For Riccioli, who speaks here for all Aristotelian geocentrists, the physical evidence is not that of a few sensations and experiences, "but [arises] from the sensation of all [and has been] repeated nearly an infinite number of times and which maintains its force as long as the contrary does not prevail." For surely, Riccioli concludes, "if it is not evident to the sense that heavy bodies descend through a straight line, nothing will be evident to it and the whole of physical science will be destroyed. . . ."

Riccioli also appealed to intuition. He argued—as did virtually all Aristotelians—that "the nature of heavy and light bodies demands that they be returned to their places and united to their whole by means of the shortest path."[148] On the assumption of the earth's rotation, however, the paths of heavy and light bodies would be curvilinear and longer, rather than perpendicular and shorter. With a seeming sense of contempt, Riccioli accuses the Copernicans of saving their hypothesis at any cost, even ignoring the nature of heavy bodies.[149]

---

[147] Ibid., 473, col. 1, sextum argumentum (the sixth argument).

[148] Ibid., 473, col. 2, septimum argumentum (the seventh argument).

[149] In a brief, related argument (ibid., 475, col. 2, trigesimumsextum argumentum [the thirty-sixth argument], Riccioli appears more evenhanded. After declaring that "The Copernican motion of the earth removes the simple up and down motion of things from the universe, therefore it must not be admitted," he explains that "the Copernicans reply by denying the antecedent with respect to apparent motion," that is, the rectilinear motions we observe are

The relationship between a moving ship and objects dropped from its mast or hurled upward from its deck did not form part of the traditional core of arguments about a moving earth and the objects that moved up and down with respect to its surface. Such arguments were not, however, unknown in the Middle Ages. In the fourteenth century, Nicole Oresme had declared that if a man "drew his hand in a straight line down along the ship's mast, it would seem to him that his hand were moving with a rectilinear motion"[150] even though that motion is the resultant of two distinct motions, vertical (the hand) and horizontal (the ship). Nevertheless, arguments involving relative and compound motions did not become a regular feature of the controversy over terrestrial rotation until after Copernicus utilized them in defense of his own position.[151]

We saw earlier that Mastrius and Bellutus cited Galileo's claim that a stone dropped from the top of a mast on a moving ship would fall to the foot of the mast, and not into the sea, because the stone shares the ship's motion.[152] To refute Galileo, Mastrius and Bellutus appealed to Johannes Cottunius (1577–1658), a professor of philosophy and theology, who, in a commentary on Aristotle's *Meteorology* (Bk. 1, lecture 16) claimed that he had witnessed the fall of stones from the masts of ships and not once did any of them fall to the foot of the mast; rather they dropped into the water off the stern of the ship.[153] Admitting that they had never observed such a demonstration, Mastrius and Bellutus were nevertheless convinced that reason (*ratio*) would yield the same result. They argued that if the earth actually rotated, two motions should be distinguishable in a ship moving eastward: (1) the common west to east motion of the earth, and (2) the eastward motion caused by the force of the wind. Although the ship is influenced by both motions, the stone, when dropped, would be

---

only apparent. In the *Dialogue Concerning the Two Chief World Systems* (Drake trans., 167), Sagredo asserts that if the earth rotates, "straight motion goes entirely out the window and nature never makes any use of it at all," to which Salviati assents.

[150] *Le Livre du ciel et du monde,* Bk. 2, ch. 25, 525 of Menut's edition and translation; see also Grant, *Source Book in Medieval Science,* 505.

[151] See Rosen (tr.), *Nicholas Copernicus On the Revolutions,* Bk. 1, ch. 8, p. 16.

[152] By analogy, Galileo held that a body falling from a tower would fall at the foot of the tower because tower and stone share the earth's common rotation.

[153] Cottunius was a Greek who studied at the Greek College in Rome and even founded a college for indigent Greeks at Padua in 1653. In addition to philosophy and theology, he also earned a doctorate in medicine at Padua. His commentary on the *Meteorology* was apparently unpublished. See Charles H. Lohr, "Renaissance Latin Aristotle Commentaries: Authors C," *Renaissance Quarterly,* 28, nr. 4 (winter 1975): 724–725. Mastrius and Bellutus were presumably aware that Galileo claimed the opposite when he implied that he had himself carried out the experiment and found that a stone or ball would indeed always fall to the foot of the mast on a moving ship (see *Dialogue Concerning the Two Chief World Systems,* The Second Day, 144 [Drake trans.]; Galileo presents the typical Aristotelian interpretation of the ship experiment on 126; see also 180, where Salviati declares that the anti-Copernicans have never dropped a body from the mast of a moving ship).

influenced only by the earth's rotation and not by the ship's own eastward motion. Under these circumstances, Mastrius and Bellutus concluded that a stone dropped from the top of the mast would fall rectilinearly but would not terminate its motion at the foot of the mast. Presumably it would hit the deck somewhat to the west of the mast as the ship moved eastward during the time of the stone's fall. Thus, while Mastrius and Bellutus assumed for the sake of argument that ship and stone would share in the earth's daily rotation, they denied that the falling stone shared the ship's eastward motion. With its vertical fall independent of the ship's eastward motion, the stone would necessarily fall to the west of the mast.

On the assumption that the ship sails westward, however, the westward motion would serve to retard the ship's eastward motion as the earth rotated from west to east. Once again, Mastrius and Bellutus proceed on the supposition that the falling stone is affected only by the earth's common eastward motion and not by the ship's proper motion. Since ship and stone share in the earth's common eastward rotation and the ship's eastward motion is retarded by its actual westward course, it follows that the stone will move eastward with a velocity greater than that of the ship and, as a consequence, will fall into the water off the stern.

Galileo took a quite different approach. For him, the illustration of a ball dropped from the mast of a moving ship was only intended as an analogy with a ball dropped from a height to the surface of the rotating earth. Just as ball and ship share a common horizontal motion, so also do earth and ball share a common horizontal, circular motion. Galileo did not also apply the common motion arising from the earth's rotation to the ship argument. By contrast, Mastrius and Bellutus conflated the two distinct examples. They linked the earth's common motion with the falling stone but divorced the stone's motion from that of the ship.

In a similar manner, Amicus argued that an arrow shot upward from the deck of a moving ship would not return to the place from whence it was launched.[154] Like Mastrius and Bellutus, Amicus assumed that the arrow's path was independent of the ship's motion, from which he concluded that the greater the velocity of the ship the farther behind it would the arrow fall. From such arguments, scholastics like Mastrius, Bellutus, and Amicus[155] convinced themselves that the earth did not rotate.

3. Cannon Balls to East and West

During the Middle Ages, the relationship to a rotating earth of cannon balls fired in opposite directions was an unheard of problem, but was

---

[154] Clavius (*In Sphaeram Ioannis de Sacro Bosco Commentarius,* 213) was probably the source of this argument.

[155] Riccioli (*Almagestum novum,* pars posterior, 419, col. 1 and 428, col. 1) made only brief mentions of the fall of bodies from the masts of moving ships.

thrust into the Copernican controversy by Tycho Brahe. In letters written to Christoph Rothmann (*d. ca.* 1608) between 1586 and 1590, Tycho denied the Copernican claim that a heavy body falls simultaneously with rectilinear and circular motions.[156] Because the two motions would be natural to the body, Tycho concluded that they would interfere with each other. Moreover, how could bodies that fell with a variety of rectilinear speeds move with the same rotational speed as the earth? Indeed even if one conceded that a body detached from the earth's surface could somehow move with two such simultaneous motions and thus follow the earth's rotation, a third and violent motion, which would render the rotational hypothesis untenable, also had to be considered. Tycho imagined that a lead, iron, or stone ball was fired first toward the east after which, from the same location, another ball of equal size and weight would be fired toward the west. Each cannon ball would be moved by three simultaneous motions: (1) a natural motion toward the earth's center; (2) a natural rotational motion following the earth; and (3) a violent motion caused by the powder exploding in the cannon. Convinced that the natural, downward motion of a projectile hurled upward cannot commence until the violent upward motion is destroyed, Tycho applied this reasoning to the cannon balls. If they possessed a natural rotational motion transmitted to them by a really rotating earth, that natural, circular motion would be impeded by the violent motion caused by the powder exploding in the cannon. Consequently, the ball fired eastward should advance hardly any distance from the cannon because the latter will be carried swiftly eastward with the rotating earth, while the cannonball will move only with its violent eastward motion. The two eastward motions would prevent much of a separation. By contrast, the cannonball shot westward should be far removed from the cannon because the latter will be carried eastward by the rotating earth while the cannonball moves westward by virtue of its violent motion, which also negates its natural circular motion. Experience reveals no such discrepancies but shows rather that the cannonballs would travel equal distances.

Because of Tycho's great prestige, his argument should have served the cause of the traditionalists.[157] It was not, however, widely cited by scholastics, perhaps because it was more complicated than many others that could be invoked. Mastrius and Bellutus, however, furnished a variant of

---

[156] The letters are contained in his *Epistolarum astronomicarum liber primus* which was published at Uraniborg in 1596, Nuremberg in 1601 and Frankfort in 1610. The letters are reprinted in *Tychonis Brahe Dani Opera Omnia*, vol. 6 (1919), edited by J. L. E. Dreyer, 218–223. My discussion of Tycho is drawn from Ronald J. Overmann, "Theories of Gravity in the Seventeenth Century," 11–15.

[157] It was an argument that Galileo attempted to meet in a number of places (see *Dialogue Concerning the Two Chief World Systems*, 126–127 [where north-south shots are also considered], 168, 171, 174, 180).

the argument when they declared that a cannonball shot toward the west should have a greater impact than one shot toward the east. The earth's easterly rotation will cause the greater and lesser impacts. Thus if the cannonball fired westward struck a house, the latter, carried eastward by the earth's swift rotation, would meet the cannonball head on and, as Mastrius and Bellutus put it, "the impetus toward the west would be as if doubled."[158] Conditions toward the east are radically different. Here the force of impact is diminished because the house is moving away from the oncoming cannonball.

Arguments about cannonballs fired toward the cardinal points were of considerable interest to Riccioli. Included among his examples was one that was contrary to what Mastrius and Bellutus had proposed. In this example, Riccioli declared that

if the earth were moved with a daily motion, or even an annual motion, the same ball that is thrust forward by the same force for the same distance once to an eastern target and then to a western target would strike the eastern target with a stronger impact than the western target.[159]

As was common in such arguments, Riccioli appealed to experience: such effects were not perceived, as could be illustrated with ivory balls on a gaming table. If an immobile ivory ball were placed in the middle and then struck by another ivory ball first from the west and then from the east over the same distance and with the same force, the impact on the immobile ball would be visibly the same. Moreover, a rotating earth should affect the impetus which a ball possesses as it is projected toward the east or west. A ball projected eastward would be aided by the earth's eastward motion, which would add to the impetus imparted by the cannon or projector. By contrast, a ball hurled or projected westward would be affected by two oppositely directed impetuses: the impetus driving the ball westward would be retarded by the impulse of the ball to follow the earth's rotation eastward; and, in a similar manner, the impetus that would normally carry the ball eastward with the earth's rotation would be diminished, or interfered with, by the contrary impetus impelling the ball westward. In short, for westward cannon shots, the two impetuses resist and interfere with each other; for eastward shots they reinforce each other.[160]

---

[158] ". . . quia versus occasum veluti duplicaretur impetus . . ." Mastrius and Bellutus, *De coelo*, 562, col. 1, par. 113.

[159] Riccioli, *Almagestum novum*, pars posterior, 427, col. 2.

[160] As a specific illustration of how impressed forces could interfere with, or reinforce, each other, Riccioli considered (ibid., 428, col. 1) the impact of a one ounce clay ball in two different situations. The first involved a clay ball shot downward from a tower onto a muddy target 30 feet away. The penetration of the clay ball is deep because two impetuses reinforce each other, namely, that which was impressed by the engine that hurled it downward and the impetus produced by the heaviness (*gravitas*) of the clay ball itself. But if the same engine projected the same one ounce clay ball upward the same distance of 30 feet against the

The argument just described was the ninth of fourteen that Riccioli formulated against the daily motion of the earth based on "the motion of elementary bodies toward the four cardinal points of the world."[161] The fifth through eighth arguments also involved cannonballs fired toward two or more of the cardinal points. The fifth[162] was but a summary of Tycho's argument about the conflicting, simultaneous motive forces operating in a cannonball shot with equal force to the east and then to the west.[163] Tycho's conclusion formed the basis of Riccioli's syllogistic argument:

If the earth were committed to a daily rotation, one and the same cannonball fired in the same way would traverse less distance to the east than when fired to the west. But one and the same cannonball fired in the same way [east and west] should not traverse less distance in the east than in the west. Therefore the earth is not committed to a daily rotation.

Riccioli observes that both William Gilbert and Kepler disagreed with Tycho's analysis.[164] In his *Epitome Astronomiae Copernicanae*, Kepler had assumed a cannonball shot to the west with a force sufficient to traverse one German mile in two minutes during which time the earth rotated eastward eight miles. Since the cannonball possessed both motions, it must have been carried eastward even as it was moving westward. Indeed its net motion was wholly eastward because, as Kepler concluded, the cannonball was ultimately carried seven miles to the east, even though it fell only one mile west of the place from which it was shot. Similarly, if the cannonball were shot with a force sufficient to project it one mile to the east, that one mile must now be added to the eight miles it traverses as a consequence of the earth's eastward rotation. Thus the cannonball will

---

same target of equally soft mud, the impetus produced by the body's heaviness, or gravity, would hinder the impetus impressed by the engine that projected it upward: the action of the former impetus would tend to carry the clay ball downward and thereby resist the upward impetus impressed in that same ball by the engine. On the assumption that the nature of things was preserved, Riccioli believed there was no solid and persuasive response to this argument ("Cui sane argumento non invenio responsionem solidam et salvis rerum naturis persuasibilem").

[161] *Almagestum novum*, pars posterior, ch. 21, 423, col. 1. The fourteen arguments occupy pages 423, col. 1-429, col. 1. For a reference to the first of these arguments, see above, 43–44 and n. 145. The essence of the ninth argument was later presented by Riccioli as the sixteenth of thirty-eight summary arguments in defense of the earth's immobility (see **474**, col. 1).

[162] Ibid., 424, col. 2-425, col. 2.

[163] See above, p. 43. As he so often did, Riccioli quoted the text, in this case Tycho's *Epistolarum astronomicarum liber primus,* 189.

[164] Riccioli quotes passages from Gilbert's *De magnete,* Bk. 6 and Kepler's *Epitome Astronomiae Copernicanae*, Bk. 1, 139. In the course of the fifth argument, he also refers to an unspecified treatise by one Petrus Herigonius.

have traversed a total of nine miles. The net effect, however, is to locate the cannonball only one mile east of the place from which it was shot.

In Riccioli's judgment, Kepler failed to meet Tycho's argument. Tycho's intent was to deny that the distance of the cannonball from its original terminus was the net result of a simple subtraction or addition of the distances traversed by each of the cannonball's motions taken singly or in succession. It was his belief, according to Riccioli, that the final location of the cannonball would be the result of a mutual interaction of the motive forces within the cannonball. On this assumption, a cannonball shot westward with a force sufficient to impel it one mile during the time the earth rotated eight miles to the east, would not, as Kepler argued, travel a net total of seven miles eastward as it landed in a place one mile west of the point from which it was fired. Rather, as Riccioli interpreted Tycho, if the earth really rotates, the two contrary motions of the cannonball ought to interfere with each other sufficiently to cause the cannonball to fall noticeably short of its westward terminus of one mile and to fall noticeably short of the point eight miles to the east. As a consequence of the mutual interaction and obstruction of the two contrary motive forces acting on the cannonball, the latter would fall approximately one-fourth of a mile short of its westward terminus (a result attributable to the contrary eastward daily rotation) and approximately one-fourth of a mile short of its eastward terminus (a result attributable to the contrary westward motion of the cannonball caused by the impetus imparted to the ball from the explosion of the gunpowder), which lies eight miles east of the point from which the cannonball was fired. Because such discrepancies had never been detected, Riccioli, following Tycho, denied the earth's daily rotation.

Riccioli's sixth argument[165] is but a variation on the theme of the fifth argument and, as with the fifth, is drawn directly from Tycho Brahe. This time, the variations in distance occur when cannonballs are fired north and south, that is, along meridians toward or away from the poles. Riccioli argues that if the earth rotates and a ball were shot toward the poles along the plane of a meridian, the daily motion would cause a smaller difference in the distance traversed than when the ball is shot either to the east or to the west. Moreover if it were shot on parallels near the poles, where the earth's rotation would be slowest, the distance traversed in a given time would differ—it would presumably be less than—from the distance traversed if the cannonball were fired toward the poles from parallels near the equator. Experience, however, shows no such discrepancies from which Riccioli concluded, as did Tycho, that the earth does not rotate.[166]

---

[165] *Almagestum novum*, pars posterior, 425, col. 2-426, col. 1.
[166] Riccioli summarized this argument later on 474, col. 1 (the seventeenth argument).

Riccioli's seventh argument was one that Galileo discussed and refuted, apparently to Riccioli's satisfaction. If the earth rotated, a cannonball fired to the east or west would fail to hit its target, but would fall above or below. Since no such effects are observed, one must conclude that the earth does not rotate.[167] Except for the enunciation of the argument, the entire discussion is devoted to Galileo's analysis of the problem.[168] As Galileo described the argument, which Riccioli summarizes, if tangents are taken to the eastern and western horizons, the stars in the east appear to rise as the eastern parts of the earth drop below that tangent, while the stars in the west appear to go down as the western parts of the earth seem to rise. "Hence the shots which are aimed along this tangent toward an eastern target (which is going down while the ball is traveling along that tangent) ought to arrive high; and those to the west, low, because of the rising of the target while the ball goes along the tangent."[169] Galileo's response, as Riccioli reports, is to observe that as the earth rises above the tangent in the west, the cannon would also rise above it and new tangents would be elevated that would maintain the same relationship to the rising target.[170] As the second response, Riccioli reports Galileo's reply that no one seems to have tested the claims that the cannonball would rise above or fall below its eastern and western targets. He repeats Galileo's calculations based on an imaginary cannon shot of 500 cubits (or yards) westward along the equator. Despite the earth's eastward rotation, Galileo showed that "the error of the ball because of the diurnal motion of the earth does not exceed 4/100 of a cubit, or about one digit in width," that is, approximately one inch. Since such small differences were undetectable by any means available in the seventeenth century, Riccioli seemingly agreed with Galileo that "whether the earth rests or is moved cannot be demonstrated by such an experiment."[171] Thus did Riccioli once again neutralize

---

[167] *Almagestum novum*, pars posterior, 426, cols. 1–2. The seventh argument is later summarized as the eighteenth of thirty-eight in defense of the earth's immobility (see 474, cols. 1–2).

[168] For Galileo's discussion, see *Dialogue Concerning the Two Chief World Systems*, The Second Day (Drake trans.), 180–182. The quotations below are also from Drake's translation.

[169] Galileo, *Dialogue*, 180.

[170] Actually Galileo speaks only of the eastern horizon declaring that "just as the eastern target is continually setting because of the motion of the earth under a motionless tangent, so also the cannon for the same reason continually declines and keeps on pointing at the same mark so that the shots carry true" (180). Drake explains that Galileo here applied his erroneous theory of circular inertia.

[171] The two quotations from Riccioli actually form the conclusion of a single sentence: ". . . et error globi ob diurnum terrae motum non excederet quatuor centesimas unius cubiti, seu unum fere digitum in latum; nec posse tali experimento convinci quiescatne, an moveatur Terra." *Almagestum novum*, pars posterior, 426, col. 2. Riccioli's only criticism of Galileo concerns an alleged mistake in which Galileo "confused the chord with the sine of the arc of one minute, for the sine of such parts is 29, the radius is 100,000, so that the chord of such parts is not 30, but 58."

another of the numerous arguments he had apparently formulated in favor of the earth's immobility.

Indeed in the eighth argument, Riccioli admitted that arguments against the earth's rotation involving comparison of distances traversed were hardly obvious. More telling were those that considered the impact of bodies and the impetuses that produced those impacts.[172] As evidence of this, Riccioli proposed an example that is depicted in the figure below, which he also supplied.

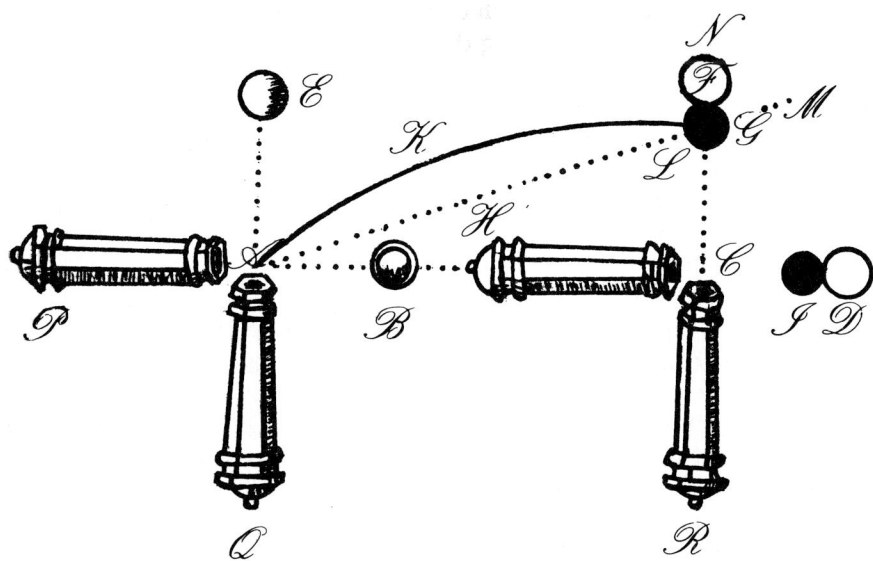

One cannonball will be fired to the east and another to the north. Identical conditions are assumed for both shots—that is, the same cannonball will be used for both, the same quantity and quality of powder, and so on. If the earth were stationary, a cannonball shot from cannon PA would reach its target B, 250 feet away, in two seconds of time. But since the earth and the bodies near and on its surface are assumed to rotate with a daily motion, the earth at the equator, from which the shot is presumably fired, will have moved 752 feet in those two seconds[173] and carried the cannon

---

[172] "Quamdiu spatia apparentia, quae a corporibus pertransiri solent spectamus, argumenta inde contra Telluris motum sumpta, non habent vim adeo manifestam; at si percussionis et impetus realis incrementum consideramus, aliquanto validiora inde tela nobis suppeditantur contra ipsius motum, ut ex dictis cap. 19." Ibid., 426, col. 2.

[173] Riccioli speaks of 752 Roman, Geometric paces (*passus*) and draws upon an earlier table (see 415) in which he gives equivalent units from different systems. I shall speak only of

to point *C*. During the same time, target *B* has also moved 752 paces to the east and reached point *D*. Thus at the end of two seconds, the cannonball will have reached its target at *D*, 250 feet away from the cannon, which is then at *C*. Riccioli next assumes that cannon *PA* is turned in direction *AQ* to fire northward at target *E*. Here again, if the earth rested, the cannonball would move to *E* over the rectilinear path *AE*, which equals *AB*, or 250 feet. Because of the earth's motion, however, and that of all the bodies surrounding it, the cannon, *AQ*, and the target *E*, will be carried 752 feet eastward to positions *CR* and *N*, respectively. The cannonball fired while the cannon was at *AQ* will now be at *F*, where its distance from the mouth of the cannon, *CF*, equals 250 feet. But in the world space, the distance traversed by the cannonball from the mouth of the cannon at *A* to its target at *N* will be more than 752 feet. For it will not have followed along the path *AC*, which is 752 feet, but along path *AKF*, whose chord, *AHL*, will be 825 feet, a figure Riccioli derived from the laws of triangles (*ex legibus Triangulorum*) applied to triangle *ACF*, where *C* is a right angle, *AC* equals 752 feet, and *CF* is 250 feet. Because the cannonball traveled only 752 feet when it was fired eastward and struck target *D*, but traveled 825 feet when it was fired northward and struck target *N*, Riccioli concluded that the cannonball struck *N* with a weaker impact than it struck *D*, a difference that would be observable if the target was a wall or another ball.

But if the earth rotates what could cause a cannonball to have a lesser impact when shot at a target directly north than when it is shot from the same cannon at the same target located in the east or west? Riccioli believed that two causes working concurrently would weaken the impetus of a ball shot northward. The first cause derives from the fact that the ball shot northward must, because of the earth's rotation, travel farther (825 feet) than when it is shot eastward (752 feet). The greater distance in the former case results directly from the earth's rotation which forces the cannonball eastward as it heads northward toward target *E*. Thus, instead of a rectilinear path over line *AE*, the ball is carried in a curved path over line *AKF* toward *F*. That the path is curved line *AKF*, rather than straight line *AHF*, occurs, according to Riccioli, "because in the beginning of the motion, the motion is quicker and the ball is carried beyond straight line AHF which it would describe if the motion were uniform."[174] If the earth rotates, the impetus of the cannonball would be diminished in the northward shot because

---

feet and make no distinction between *passus* (paces) and *pedes* (feet) since these are often used synonymously. In any event, the units are not germane to the argument. During the two seconds in which the cannon ball moves to its target, the earth's equator has rotated 30″, or 752 feet.

[174] ". . . quia in principio motus hic velocior est et globus fertur ultra rectam *AHF*, quam describeret si motus esset uniformis." *Almagestum novum*, pars posterior, 427, col. 1.

the cannonball must traverse a greater distance to reach a target 250 feet away than it does to strike the same target 250 feet to the east.

But a second cause also serves to weaken the impetus of a northward shot. To an observer at point C, it would appear that the cannonball would strike target N in point F along line FN. But actually, says Riccioli, the cannonball would strike N at F obliquely according to line LM. The earth's daily rotation deflects the cannonball, and therefore its impetus, from paths AE and FN to AHL, which is but a section (*portio*) of the curved path of the cannonball, and AKF.[175] Thus only if the target were moved from N in point F to point G would there be a greater impetus and impact.

From the evidence just presented, Riccioli generalized his eighth argument: "If the earth were moved in a diurnal, or also in an annual, motion, the impact of a cannonball shot towards the North or the South would be much weaker than that [of the ball shot] from the West to the East."[176] In the absence of such observed differences, Riccioli concluded that the earth does not rotate.

In anticipating how Copernicans might reply to his arguments, Riccioli may have had Galileo in mind.[177] Despite its oblique path, Copernicans insisted that a cannonball had its own proper motion and struck a target directly because target, cannon, and cannonball are all traveling with the common motion of the earth. Copernicans were committed to an interpretation that demanded they analyze every terrestrial motion as if it were compounded of two motions, its own proper motion and the common motion that it shared with the earth and all other objects. The two component motions did not, however, interfere with each other, an interpretation that Riccioli could not accept because of his conviction that each motion of a body supplied a quantity of impetus to it. If two or more distinct motions were actually operative in the production of an observable motion, the impressed forces associated with those motions must necessarily interfere with each other. Such mutual interference was not confined to contrary forces, but also occurred with forces that were impressed

---

[175] It is unclear why Riccioli speaks of AHL as the path of the cannonball when earlier he explained that AHL would be the path only if the motion were uniform. Since the motion is not uniform, its path ought to be the curved line AKF.

[176] Translation by Alexandre Koyré, "A Documentary History of the Problem of Fall from Kepler to Newton, *De Motu Gravium Naturaliter Cadentium in Hypothesi Terrae Motae*," *Transactions of the American Philosophical Society*, New Series, Vol. 45, part 4 (1955), 355, n. 132. Koyré's translation is based on Riccioli's Latin text (which he also quotes on 355) as it was reprinted in James Gregory's report of a controversy between Stefan degli Angeli and Riccioli on the fall of bodies on an earth that was assumed to rotate. Gregory's version of Riccioli's Latin text, which appears in the *Philosophical Transactions of the Royal Society*, 1 (1688): 684, differs in inconsequential ways from Riccioli's text in the *Almagestum novum*, pars posterior, 427, col. 2.

[177] Galileo, whom Riccioli did not explicitly cite in this context, discussed vertical shots of cannonballs in the *Dialogue Concerning the Two Chief World Systems*, The Second Day (Drake's trans.) 175–179.

obliquely.[178] With his assumption that impetus was supplied to a cannonball by both the powder that was exploded in the cannon *and* the earth's eastward rotation,[179] Riccioli was committed to a wholly different analysis of motion.

4. *The Fall of Heavy and Light Bodies*

Relatively few scholastics had the technical competence to cope with arguments for or against the earth's rotation that derived from the assumption that bodies fell with uniformly accelerated motion. Because of his proficiency in astronomy and mathematics, Riccioli was a notable exception.[180] According to Riccioli, Copernicans had determined that if the earth rotated with a daily motion and also moved with an annual motion, heavy bodies could not fall in a straight line perpendicular to the horizon but would fall with either a parabolic (Kepler and Gassendi) or circular (Galileo and Bullialdus) motion.[181] Toward the refutation of these claims, Riccioli devoted considerable space in the *Almagestum novum* and in later works. Within the *Almagestum novum* itself, Riccioli considered the problem in a number of places. Those who, like Bullialdus and Galileo, assumed the earth's rotation explained the apparent downward rectilinear path of a heavy body dropped from a tower as the composite of two distinct motions. Thus Bullialdus distinguished two uniform circular motions, whereas Galileo identified a common uniform circular motion and the body's own uniformly accelerated rectilinear motion.[182] Indeed, upon further analysis, Galileo had Salviati declare that "the true and real motion of the stone is never accelerated at all, but is always equable and uniform"[183] as it moves along its circumferential path.

Galileo's explanation made little sense to Riccioli because its truth implied that heavy bodies "would fall from a more elevated place with no greater

---

[178] "Certi enim sumus," Riccioli insisted, "ex plurimis experimentis motum semel impressum ac motivum versus unam partem debilitari ac minui ab impetu non tantum in contrariam sed etiam in alienam partem, seu in transversum movente." *Almagestum novum,* pars posterior, 427, col. 2.

[179] See above, 51–52.

[180] Riccioli himself had provided one of the first careful experimental determinations of acceleration. For a description and analysis of Riccioli's experimental work, see Alexandre Koyré, "An Experiment in Measurement," *Proceedings of the American Philosophical Society,* 97, part 2 (1953): 222–237. See also, Koyré, "A Documentary History of the Problem of Fall," *Transactions of the American Philosophical Society,* 45, part 4: 349.

[181] For corrections to Riccioli's claims about Kepler and Gassendi, see Koyré, "A Documentary History of the Problem of Fall," 349, n. 91. For Galileo's discussion in the *Dialogue Concerning the Two Chief World Systems,* which was of primary concern to Riccioli, see 164–167 of Drake's translation.

[182] Riccioli's description and refutation of the arguments of Bullialdus and Galileo in *Almagestum novum,* pars posterior, Bk. 9, sect. 4, ch. 17, pp. 398–401 have, with some omissions, been translated and annotated by Koyré, "A Documentary History of the Problem of Fall," 349–354.

[183] *Dialogue Concerning the Two Chief World Systems,* The Second Day, 166 (Drake trans.).

impetus than from a lower one, and therefore innumerable effects which result from the more vehement percussion of [bodies] falling from a higher place . . . would not occur."[184] Riccioli believed that his own experiments had demonstrated that "if two heavy bodies of different weight are dropped at the same time from the same height, that one which is heavier will descend more quickly, if it is heavier both individually and specifically . . ."[185] According to Galileo, however, those heavy bodies ought to strike the ground at the same time.[186] But even if Galileo's explanation were correct, Riccioli was convinced that it did not "represent the greatest part of the motions by which heavy bodies descend naturally . . ."[187] Galileo's circular motion hypothesis, derived from the example of a stone dropped from the top of a tower, was only applicable to bodies falling directly over the equator, for, as Riccioli explains, "if it were outside of it, under the poles, the descent of the stone would in fact be on a straight line and therefore not on a circle; if, on the other hand, [it were] on some parallel of the Equator, the parallel described by the foot of the tower would be different from the [one] described by its summit; and besides, the plane of neither would be in the plane in which is the center of the Earth, but in a quite different one."[188]

Later in the *Almagestum novum*, Riccioli applied to the problem of the earth's rotation Galileo's distance and time formulation for falling bodies

---

[184] Koyré, "A Documentary History of the Problem of Fall," 352, col. 2. "This conclusion of Riccioli," Koyré declares, "is, of course, completely erroneous, and it is based on his inability to interpret correctly the meaning of Galileo's and Bullialdus's theory . . ." (ibid., n. 108). Nevertheless, Koyré considered this to be Riccioli's strongest argument (ibid., 354).

[185] Koyré's translation, ibid., 352, col. 2.

[186] Riccioli remarks (Koyré's trans. ibid., 353, col. 1) that Father Grimaldi had hypothesized that Galileo had denied that two unequal heavy bodies dropped simultaneously from the same height would strike the ground at different times because it would contradict his claim that their motions were uniform. Grimaldi does, however, conjecture that perhaps Galileo "observed two globes of different weight and bulk but of the same kind (species); in this case the difference in the descent and in the percussion appears much smaller than in the other comparisons; and does not manifest itself evidently if [they are] not released from a very great height. But, as it is attested in the same Dialogue, Galileo did not make use of an altitude greater than 100 cubits." Koyré, however, observes (ibid., 353, n. 116) that "Father Grimaldi is in error; Galileo knew perfectly well that on the earth, *in hoc vero aere*, heavy bodies fall more quickly than light ones."

[187] Koyré's trans., ibid., 353, col. 2.

[188] Koyré's trans., ibid., 353–354 (the bracketed additions are Koyré's). In the *Dialogue Concerning the Two Chief World Systems*, Galileo had Simplicio make a similar criticism (see p. 219 of Drake's translation) drawn from a "booklet of scientific theses" (ibid., 218), which Drake identifies as the *Disquisitiones mathematicae de controversiis ac novitatibus astronomicis* (Ingolstadt, 1614), "a book written at the instigation of [Christopher] Scheiner by his pupil [Johann Georg] Locher" (ibid., 476). The bracketed additions are mine. Indeed Riccioli seems to have derived this argument from the *Disquisitiones*, which he ostensibly quoted (Koyré, "A History of the Problem of Fall," 354). Koyré, however, who includes Riccioli's Latin quotation from the *Disquisitiones*, declares emphatically that "since Galileo's time no one, probably not even Father Mersenne or Father Riccioli, though they both quote them, has ever so much as looked at the *Disquisitions*" (ibid., 331). Koyré believed that Riccioli borrowed his quotations from Galileo's *Dialogue*.

(that is, s $\propto$ t^2), which he (Riccioli) had himself experimentally verified.[189] As his most telling argument, Riccioli declared that

heavy bodies let fall through the air in the plane of the equator descend toward the Earth with an increased speed that is real and notable, not only apparent. But if the Earth did move in a diurnal motion only around its center, no heavy body let fall through the air in the plane of the equator would descend towards the Earth with a real and notable increase of speed, but only with an apparent one. *Ergo*, the Earth either does not move, or does not move in a diurnal motion only.[190]

Balls he had dropped from different heights of a tower proved to Riccioli that not only does the speed of a body increase according to the square of the time but that the force of the impact increases as the falling body acquires more and more impetus.[191] If, however, the earth rotates, these variations in impact cannot occur because a ball dropped in the plane of the equator would describe a uniform circular path and thus "would not descend with a real inequality or with a real increment of velocity."[192]

The arguments that Riccioli formulated for the motion of heavy bodies were also intended for the natural upward motion of light bodies. Faithful to Aristotle and the centuries long Aristotelian tradition, Riccioli, in the fifth of five arguments against the diurnal rotation based on the increment of velocity of heavy and light bodies, not only reiterated his conviction that these arguments were applicable to the motion of light bodies but also conjectured about the conditions under which their application would be evident to us. The light bodies would have to be free of all earthy and watery natures and be purely airy or fiery. If a body like this existed and were visible beyond the region of our air as it ascended perpendicularly,

such a body, inasmuch as it is not cognate to the earth, ought not to follow the daily or annual motion of the earth. Therefore if the earth were moved, it [i.e. the purely airy or fiery body] would surely be left behind [or abandoned] in airy space and its ascent would not appear perpendicular to us but would appear oblique to

---

[189] See Alexandre Koyré, "An Experiment in Measurement," *Proceedings of the American Philosophical Society*, 97, pt. 2: 229–32 and "A Documentary History of the Problem of Fall," 349.

[190] The translation is Koyré's ("A History of the Problem of Fall," 355, n. 131) from the text given by James Gregory in the latter's account of the controversy between Stefan degli Angeli and Riccioli on the motion of the earth which Gregory published in the first volume of the *Philosophical Transactions* of the Royal Society. Gregory drew the Latin text from Riccioli's *Almagestum novum*, pars posterior, 409, col. 1. Although Riccioli presented many arguments against the earth's motion, Gregory singled out three that he considered the strongest, the first of which is the argument cited above. The three arguments formed the basis of Gregory's article, which Koyré reproduced with English translations of the Latin passages quoted by Gregory.

[191] Riccioli, *Almagestum novum*, pars posterior, 409, col. 1.

[192] "Ergo non descenderet cum reali inequalitate, seu cum reali incremento velocitatis, si nimirum Terra solo motu diurno moveatur." Riccioli, ibid., 410, col. 1. The proof that if the earth rotated, a heavy body would fall with a uniform, circular motion appears on 409–410. On pages 412–413, Riccioli applies the same argument to heavy bodies that do not fall on the equator.

the west. And, conversely, if it did appear perpendicular, this would be a sign that the earth and the part of it where the observer is remain unmoved.[193]

Indeed even if we cannot observe the ascent of such a light body, it is nevertheless probable that airy corpuscles and exhalations would behave in the manner described and thus the immobility of the earth is more probable than its mobility.

5. Miscellaneous Physical Arguments

Many other physical arguments, with their seemingly endless variations, could be added to those already described. Of these, only a few of the more significant will be included. One that was widely discussed in the sixteenth and seventeenth centuries is traceable to Copernicus, who, in *De revolutionibus*, Bk. 1, ch. 7, falsely ascribed to Ptolemy the opinion that if the earth rotated "living creatures and any other loose weights would by no means remain unshaken."[194] As in so many instances, it was Clavius who installed this argument in the scholastic repertoire against a rotating earth. To those who countered—as did Galileo some years later[195]—that the earth's rotation would no more cause buildings to collapse than would the swift revolution of a vessel filled with water cause the water to be expelled, Clavius devised a response.[196] "The whole impetus of the water," he explained, "is impressed toward the lower parts of the vessel, not toward its orifice. But the impetus impressed on the buildings is toward the farthest parts of the earth."[197] In these rather cryptic words, Clavius seems to say that the water remains in the vessel because the impetus, or force, impressed on the water is totally concentrated at the bottom of the vessel so that the water tends toward the bottom of the vessel and cannot depart. The earth's rotation, however, causes the impetus to concentrate at its surface and, perhaps like an earthquake, to crumble the foundations of the buildings on it.[198] To Clavius's argument, Bartholomew Amicus, who substantially repeated it, added an important qualification, namely,

---

[193] Riccioli, ibid., 417, col. 2.

[194] *Nicholas Copernicus On the Revolutions* (Rosen trans.), p. 15. On the falsity of the attribution to Ptolemy, see Rosen's note to p. 15, line 17 on 351 and also Stillman Drake, trans., *Galileo, Dialogue Concerning the Two Chief World Systems*, 481–482 (note to p. 188). Although Ptolemy did hold the opinion ascribed to him by Copernicus, it was not with respect to a rotating earth but to one that fell with a downward motion like a stone or particle of earth. Galileo, who also ascribed the same argument to Ptolemy, reported it with a further embellishment by adding that "if the earth turned upon itself with great speed, rocks and animals would necessarily be thrown toward the stars . . ." (Drake trans., 188).

[195] *Dialogue* (Drake trans.), 189–190.

[196] Clavius does not make clear whether the water-filled vessel is conceived as rotating around its own axis or whether it is swung around on the end of a cord. Galileo (ibid.) assumed the latter.

[197] Clavius, *In Sphaeram Ioannis de Sacro Bosco Commentarius*, 213.

[198] By a parity of reasoning, the impetus impressed on the water at the bottom of the vessel ought to cause the bottom to collapse.

that the vessel must be conceived as in a horizontal orbit with its orifice always directed toward the center. We may infer this from his assertion that not even the quickest motion could keep the water in the vessel if the orifice were turned away from its center. Only if the orifice faces the center throughout its orbital swing will the water remain within.[199]

Impetus was invoked in yet another context in which the claim was made that because fire and air are moved circularly, so also ought the globe of earth be moved circularly, presumably by an impetus transmitted from heaven to earth via the spheres of fire and air.[200] Mastrius and Bellutus denied the physical feasibility of this claim simply because a fluid body like air could not push a solid body like earth. The latter is indeed not only too heavy to be rotated by the action of the air but too distant from the heaven to be affected by any impetus transmitted by the heaven to fire or air.[201] More significant yet was Amicus's assertion that even if the buildings could stand for a time on a rotating earth, they must eventually collapse as a consequence of that rotation.[202] A few years later, Galileo insisted that those who believed buildings would collapse on a rotating earth could not also believe that the earth had always rotated, for otherwise how could the buildings have ever been constructed? Partisans of this argument had to assume that the earth was initially at rest during which time the buildings were constructed. With the commencement of rotation, however, the buildings would quickly collapse.[203]

Rather than rely solely on arguments that rejected the earth's rotation, positive reasons were also offered in defense of the earth's immobility. After rejecting five alleged causes for the earth's immobility, Bartholomew Amicus sided with Aristotle in holding that the earth's heaviness caused it to rest in the center,[204] which is also the lowest place in the universe.[205] Or, as Mastrius and Bellutus would have it, the earth rests in the middle

---

[199] Amicus, *De coelo*, p. 289, col. 2.

[200] The argument is reported by Mastrius and Bellutus, *De coelo*, 563, col. 2, par. 119.

[201] Ibid.

[202] For Riccioli's similar arguments, see *Almagestum novum*, pars posterior, 432, col. 2-433, col. 1. On this theme, Aversa had little to say, noting only that "the earth would be easily dissipated by so swift a motion with which it turned incessantly" (*Philosophia*, 142, col. 2).

[203] Galileo, *Dialogue on the Two Chief World Systems*, Second Day, 189. Galileo posed this argument against Ptolemy's alleged claim that buildings on a rotating earth would collapse. We have already seen that this was not Ptolemy's argument (see also n. 194 above).

[204] Amicus, *De coelo*, 601, col. 1. The five causes which Amicus rejected include (1) that which Aristotle attributed to Colophanus and Zenophanes, who held that the part of the earth opposite us is of infinite depth; (2) the idea, drawn from the *Liber de incessu animalium*, that all motions must be made around something immobile "and [since] the celestial motions are made around the earth, it ought to follow that the earth is immobile"; (3) the notion of Thales, reported by Aristotle, that the earth is supported by water lest it fall; (4) the claim that the earth rests in the middle because of the great velocity of the celestial motions; and (5), finally, the argument which Aristotle attributes to Anaximander, namely, that the earth rests in the center because it is equidistant from everything. The five causes are described on pp. 598, col. 1-601, col. 1.

[205] Amicus, ibid., 601, col. 1.

of the universe "because it is in the lowest place."[206] The earth remains motionless at the center of the world because any movement away from the center would be an ascent, which is repugnant to the earth's heaviness.[207] That heavy things always rest at the world's center was obvious to Amicus who was convinced that if a stone were dropped through a hole imagined to extend from one side of the earth's diameter to the other, "it would not be moved except to the middle and there it would naturally rest and not proceed beyond except by force. . . ."[208]

If the earth rested at the center of the world, it obviously did not move in its place and therefore did not rotate. How, then, could one explain the movement in place of animals, which are also earthy bodies? Amicus, citing Albertus Magnus with apparent approval, replied that it is appropriate for things that have understanding to be moved in their places by reason of desire. "But the earth has neither soul nor understanding,"[209] and, therefore, cannot be moved by desire. Such analogical arguments, which were commonplace during the Middle Ages, retained their appeal for scholastics in the seventeenth century.

But if the earth did not rotate in the center of the world, was it perhaps plausible to assume that it moved with small motions caused by the forces of bodies that incessantly pushed on its surface? Amicus denied all alleged evidence in support of such a claim. When chariots drawn by four horses roar by "we do not see that a basin shakes on the pavement, or that a vessel filled with water totters or turns over; nor that our feet vibrate there."[210] But what if the earth is in equilibrium around the center and a weight were pressed on one side of it? Would this not, as with steelyards and balances, depress that side of the earth and cause the other to rise? Amicus attacks the analogy. The earth is not like a scale or a balance,

---

[206] Mastrius and Bellutus, *De coelo*, 564, col. 2, par. 121.

[207] Amicus, *De coelo*, 601, col. 2. Aversa (*Philosophia*, 231, col. 2) expressed much the same opinions when he declared: "The earth cannot decide [to move] toward the heaven because then it would truly ascend, not descend; for to ascend is to withdraw from the middle, to descend is to incline toward the middle. The lowest place is the middle of the heaven; the highest place is nearest to the heaven. Therefore, since the earth holds the middle place of the heaven, it ought to rest and remain there absolutely. Moreover, with the earth possessing the greatest gravity [or heaviness], that [alone] ought to sustain it in the middle and hold it in the middle of the air as if suspended."

[208] Ibid. Amicus also invoked Scriptural passages in support of the earth's centrality and immobility. His conclusion that a stone dropped into a hole through the earth's center would come immediately to rest had already been rejected in the fourteenth century by Albert of Saxony and Nicole Oresme, who argued that the residual, uncorrupted impetus in the falling stone would cause it to proceed past the center and ascend toward the heavens. And, as Albert put it, "in so ascending, when the impetus would be spent, it would conversely descend. And in such a descent it would again acquire unto itself a certain small impetus by which it would be moved again beyond the center. When this impetus was spent, it would descend again. And so it would be moved, oscillating (*titubando*) about the center until there no longer would be any such impetus in it, and then it would come to rest." For the translation, see Clagett, *The Science of Mechanics in the Middle Ages*, 566; for Oresme's version, see 570; also 553.

[209] Ibid.

[210] Ibid., 601, col. 2-602 [mistakenly paginated 596], col. 1.

where forces act to cause risings and fallings. The earth rests at the center of the world naturally and its parts are organized around one another in a natural manner. This "natural" equilibrium is not affected by the addition of weights, however large, to any point of the earth's surface: "otherwise it would be easy to equate mountains with valleys by assuming that the mountains of the earth were dug from the valleys [of the earth]."[211] Because of its many eminences and depressions, moreover, the earth is not a perfect sphere, so that additions of heavy matter to any particular point on the earth's surface will not cause an inclination toward any part, no more than the weight of the upper air causes us to feel the pressure of the lower parts that rest on the tops of our heads. The whole earth rests because it resists the application of any motive force to any of its parts, just as a great weight, say an enormous stone wheel, resists the application of any ordinary forces and remains at rest.

Not all scholastics categorically denied any kind of motion to the earth. Raphael Aversa was convinced that the earth endured small, albeit imperceptible, rectilinear motions. In a discussion reminiscent of John Buridan's in the fourteenth century, Aversa argued[212] that various parts of the earth suffer continuous alterations that cause them to increase or decrease in weight. As a consequence of these small, but continuous, alterations in weight, the earth's center of gravity, and therefore its center, continually shifts thereby causing an incessant sequence of small rectilinear motions of the whole earth.[213] Although such movements occur continually, Aversa considered them so minimal as to be imperceptible. Indeed this "tenuous motion of the earth, which escapes our senses . . . , must be taken as if it did not exist. . . ."[214] Thus did Aversa have his motion and deny it at the same time, leaving his anti-Copernican credentials intact.

B. METAPHYSICAL ARGUMENTS: SIMPLICITY, ORDER, AND NOBILITY

The prima facie simpler operation and arrangement of the Copernican system were powerful factors in its favor. The principle of Ockham's razor

---

[211] Ibid., 602 [mistakenly paginated 596], col. 1.

[212] Aversa, *Philosophia*, 232, col. 2-234, col. 2.

[213] Ibid., 233, col. 1. Buridan presented much the same interpretation in at least two places in his *Questions on De caelo*. In Bk. 2, question 7 ("Whether the whole Earth is Habitable"), Buridan inferred the motion of the earth and the formation of mountains as a consequence of continual alterations on the earth's surface and the attendant shiftings of the earth's center of gravity (for the Latin text, see Moody [ed.], 159–60; for an English translation, see Grant, *A Source Book in Medieval Science*, 623). The second discussion occurs in Bk. 2, question 22 ("Whether the Earth always is at Rest in the Center of the Universe"), where Buridan attributed slight rectilinear motions to the earth as a consequence of continual shifts of the earth's center of gravity (for the Latin text, see Moody's edition, 231–232; for the English translation, see Marshall Clagett, *The Science of Mechanics in the Middle Ages*, 597–598 or Grant, *Source Book*, 502, 503, where Clagett's translation is reprinted).

[214] Aversa, *Philosophia*, 234, col. 2. Although the quoted passage represents Aversa's true opinion, the ellipsis replaces some 15 lines of text.

tended to support the Copernicans, although scholastics, in meeting that famous argument, sometimes quoted some version of it, as when Amicus declared that "it is vain to do with many [things] what can be done equally well with fewer."[215] In light of the force of the principle of simplicity, would it not be "easier and of less cost [or effort]," queried Riccioli, "to move the small (*pusillum*) globe of the earth than the immense machine of the heaven? Therefore God and Nature, which do what is easier, move the earth, rather than the heaven, with a daily motion."[216] Such arguments were applied equally to speed as well as weight. If, instead of the earth, the stars and planets turned daily, they would move with incredible speeds despite their enormously greater heaviness than the earth.[217]

Like many other scholastics, Riccioli was unimpressed with simplicity arguments. The great speeds of the celestial spheres are of no consequence as long as the spheres themselves were capable of enduring such motions. Nor indeed are our senses adversely affected. Ill effects are avoided because those great planetary speeds are regulated by celestial intelligences.[218] The much greater mass of the celestial spheres would pose serious problems only if the motive forces that continually moved them met more resistance than they could cope with. Even if such resistances existed, they could cause no difficulty for God or the intelligences. Although it would be easier, perhaps, for God to move the smaller earth than the larger heavenly spheres, Riccioli alludes to valid arguments (though he cites none) as to "why God and Nature do not wish to do that which seems, at first glance, easier, just as in many other matters what seems easier, or of less cost, is not followed."[219] Despite the earth's considerably smaller size than the heavens, which might indicate a greater inclination for motion, Bartholomew Amicus insisted that the earth's heaviness made it more unsuited for motion than water, which was less suited for motion than air, which, in turn, was less suited for motion than fire from which he inferred that superior celestial bodies are far better adapted for motion in their places than is the earth in its place.[220]

[215] ". . . quia frustra fit per plura, quod potest aeque bene fieri per pauciora." Amicus, *De coelo*, 288, col. 2. For equivalent statements, see Cornaeus, *Curriculum philosophiae peripatetici*, 532 and Mastrius and Bellutus, *De coelo*, 563, col. 2, par. 119. Galileo also used and expressed the principle of simplicity at least twice in the *Dialogue Concerning the Two Chief World Systems* (Drake trans., 117, 123), giving the Latin text in the second reference ("frustra fit per plura quod potest fieri per pauciora").

[216] Riccioli, *Almagestum novum*, pars posterior, 466, col. 2 (*Quintum argumentum*). See also Amicus, *De coelo*, p. 288, col. 2; Cornaeus, *Curriculum philosophiae peripatetici*, 532, 537; and Mastrius and Bellutus, *De coelo*, 563, col. 2, par. 119. For similar arguments by Buridan and Oresme in the fourteenth century, see Grant, *Source Book in Medieval Science*, pp. 501 and 509, respectively. As a Copernican, Galileo also used this argument (see Salviati's remarks in *Dialogue Concerning the Two Chief World Systems*, The Second Day, 120).

[217] See Riccioli, *Almagestum novum*, pars posterior, 467, col. 2 (*Undecimum argumentum* [eleventh argument]).

[218] Ibid., 467, col. 2.

[219] Ibid., 466, col. 2.

[220] Amicus, *De coelo*, 291, col. 2.

The traditional scholastic conviction that rest is more noble than motion was used by Copernicans in defense of a rotating earth. With rest more noble than motion, why should the imperfect earth rest while the more perfect and noble heavens rotate? Scholastics responded in a variety of ways. Amicus, for example, conceded that rest is generally more perfect because it is the goal or end of motion.[221] Under certain circumstances, however, the reverse is true, namely, when motion produces more noble effects than rest. Nature assigned motion to the heavens because the latter acts as an agent to produce such terrestrial effects as the seasons, variation of days and nights, and the distribution of influences. Since the motion of the earth alone could not produce the various astronomical aspects and conjunctions necessary to generate these causes, nature assigned rest to the earth.[222] Mastrius and Bellutus adopted a similar approach.[223] Natural rest, that is, rest that terminates motion to a natural place, is more perfect than motion toward that natural place. But motion that does not move toward a natural place in order to come to rest there, but seeks, rather, to communicate its power to inferior things is more perfect than rest. The circular celestial motions, which operate for the good of the universe and do not come to rest, belong to this category.

The nobility argument had also been employed by Copernicans to argue for the sun's, rather than the earth's, centrality. Riccioli, who reported numerous Copernican arguments favoring the sun's centrality,[224] attacked that Copernican argument which assumed that the center of the world is the most noble place and then promptly inferred that the sun, which was usually deemed nobler than the earth, must occupy it. Riccioli conceded that *in the natural order* the center is the most noble place, but not *in the supernatural order* where the most noble place is the Empyrean sphere, the highest place, whereas the lowest place, that is, the center, is the place of the damned. But even in the natural order, the sun is not in the center because "the earth, with its living things, especially rational animals, is nobler than the sun."[225] To save the earth's centrality, Riccioli was thus prepared to abandon the traditional opinion that the sun is nobler than the earth. Moreover, he also denied that the sun was the efficient cause of celestial motions, as Kepler argued, or that it could be the cause of the elements and of new phenomena. Rather it is the earth that is the ultimate

---

[221] Ibid., 604, cols. 1–2.

[222] Amicus presented much the same argument earlier on 292, col. 1.

[223] Mastrius and Bellutus, *De coelo*, 563, col. 2, par. 119. For the argument advanced by Riccioli, see *Almagestum novum*, pars posterior, 467, col. 1.

[224] See Riccioli's "29 arguments in favor of the sun's position in the center of the universe and [in favor] of the annual motion of the earth around the center of the universe simultaneously with the daily motion, and their solutions . . ." *Almagestum novum*, pars posterior, 469, col. 1. The 29 arguments extend over pages 469–472 (Arguments 21 to 49). Except for the first argument, which is cited here, the arguments are overwhelmingly astronomical rather than physical or metaphysical.

[225] Riccioli, ibid., 469, col. 1 (*Primum argumentum*).

cause of all these motions and changes because "the earth, with its human beings, is the final cause and objective of the motion of the stars."[226]

The sphericity of the earth had also served the Copernicans, many of whom argued that the earth's spherical figure was more suited for circular motion than for rest.[227] For just as the spherically shaped celestial bodies move with circular motion, so also should the spherically shaped earth.[228] Under the Copernican threat, some scholastics denied any necessary connection between sphericity and circular motion. Amicus, for example, insisted that the earth's sphericity was more appropriate for rest "because, by reason of heaviness, parts of the earth tend to the center equally; [therefore] it [the earth] ought necessarily to have, as much as it can, a spherical figure, so that all the parts of its circumference are equally distant from the center."[229] Although Mastrius and Bellutus conceded that circular motion was indeed appropriate to the spherical earth, they denied that the earth had such a motion and offered supporting reasons.[230] Rather than the earth, it was the *primum mobile*, or first movable sphere, that rotated with a daily motion. True, the primum mobile required an enormously greater velocity to complete its daily rotation than did the far smaller earth. But that greater velocity was a direct reflection of God's omnipotence and therefore produced no disastrous consequences. Indeed so admirably did the primum mobile illustrate God's power that we ought not to reject its tremendous speed in favor of the earth's more imaginable daily rotational velocity. Moreover, if the earth rotated, as the Copernicans argued, external movers would need to be multiplied almost infinitely because every stone projected upward would require two external forces: one to move it up, the other to move it along with the earth's rotation. The daily motion is, therefore, better placed in the heavens.

Also rejected by scholastics was the popular Copernican argument that circular motion is more natural to the elements, and therefore to the earth, than is rectilinear motion.[231] In Riccioli's account of this argument,[232] circular motion is said to be more appropriate to things that are in their natural places, as when earth, water, air, and fire are in their natural places. Only when a part of an element leaves its natural place does it follow a rectilinear path. But such rectilinear motions represent disorder and disorganization because those elemental bodies have departed from their natural places

---

[226] Riccioli, ibid., 469, col. 1 (*Quartum argumentum*).

[227] See Amicus, *De coelo*, 288, col. 1; Mastrius and Bellutus, *De coelo*, 563, col. 2, par. 119.

[228] Riccioli described the Copernican position somewhat differently. In his version, the daily motion should be assigned to the spherical earth rather than to the heaven of the fixed stars because the actual sphericity of the latter was uncertain (*Almagestum novum*, pars posterior, 466, col. 1 [*Primum argumentum*]).

[229] Amicus, *De coelo*, 291, col. 2-292, col. 1.

[230] *De coelo*, 563, col. 2, par. 119.

[231] Copernicus advances this argument in *De revolutionibus*, Bk. 1, ch. 8, 16–17 (Rosen trans.).

[232] Riccioli, *Almagestum novum*, pars posterior, 466, cols. 1–2.

by violent motion. Riccioli's reply denied that circular motion was an inherent (*ab intrinseco*) property of the elements. Circular motion was imposed on the elements externally (*ab extrinseco*) and could not, therefore, be characterized as "natural." Were the elements arranged absolutely in their natural places, they would be immobile, rather than tend toward circular motion, as the Copernicans assumed. But when the elements are not in that arrangement, heavy and light bodies are moved with a natural, finite rectilinear motion along a "perpendicular line [that is] always accelerated uniformly difformly toward a goal. . . ." These rectilinear motions of bodies out of their natural places agree with observed phenomena. "For these and other reasons [or causes]," declared Riccioli, "we have taught that the Peripatetic doctrine is far more solid in this than the Copernican or Galileistic [doctrines]."

Additional physical and metaphysical arguments could be described, but what has already been presented is more than sufficient to indicate the range and substance of the arguments in these categories. Numerous astronomical and theological defenses of the earth's centrality and immobility were also formulated. Indeed Riccioli alone compiled an impressive number of arguments in both areas.[233] Although it is not the purpose of this study to describe the scholastic astronomical and theological arguments in behalf of a stationary and central earth (a few astronomical arguments were, however, described earlier), a brief consideration of the essential character of the theological arguments should provide a better understanding of the continued survival and strength of Aristotelian-Ptolemaic astronomy through most of the seventeenth century.

The theological arguments consisted largely of Biblical passages that mentioned an immobile earth and/or a mobile sun circling the earth. By the seventeenth century, certain Biblical passages were regularly invoked in support of traditional geocentric cosmology. Clavius, whose influence on seventeenth century scholastic cosmology cannot be overestimated, was an immediate source for at least three of these,[234] namely, Psalms 103:5,[235] Ecclesiastes 1:4–5,[236] and Psalms 18:6–7.[237] In these passages, Clavius saw

---

[233] The astronomical arguments are found primarily in *Almagestum novum*, pars posterior, Bk. 9, section 4 (*De systemate terrae motae*), 290–478; the theological arguments in Bk. 9, section 4, 479–500.

[234] Clavius, *In Sphaeram Iohannis de Sacro Bosco Commentarius*, 214.

[235] "Who hast founded the earth upon its own bases; it shall not be moved for ever and ever." This, and all subsequent quotations are drawn from the Douay Version. The passage was also cited by Amicus, *De coelo*, 290, col. 2; Aversa, *Philosophia*, 232, col. 2; Cornaeus, *Curriculum philosophiae peripateticae*, 535; and Riccioli, *Almagestum novum*, pars posterior, 480, col. 2.

[236] "4. *One* generation passeth away and *another* generation cometh; but the earth standeth for ever.

5. The sun riseth, and goeth down, and returneth to his place: and there rising again, . . ." See also, Amicus, ibid., 290, col. 2; Aversa, ibid., 5, col. 2, 142, col. 1, 232, col. 2; Cornaeus, ibid., 536; and Riccioli, ibid., 480, cols. 1–2.

[237] "6. He hath set his tabernacle in the sun: and he, as a bridegroom coming out of his bride chamber, Hath rejoiced as a giant to run the way.

sacred support for the earth's immobility and for the movement of the sun and stars. "Could anything be said clearer," he exclaimed.

Other Biblical passages were also invoked to uphold the traditional geocentric world. To confirm the claim that the earth continually supports itself in the center of the world, Raphael Aversa appealed not only to Psalms 103, but also to Job 26:7, where God suspended the world over nothing, and Isaiah 40:12, where God is said to have "poised with three fingers the bulk of the earth, and weighed the mountains in scales, and the hills in a balance."[238] As evidence of the earth's immobility, Aversa cited[239] 1 Chronicles 16:30, where God is said to have made the orb immobile,[240] and Psalms 92, which declares that God fixed the orb of the earth so that it does not move.[241] Amicus, and others, found support for the earth's immobility in 2 Kings 20:9–11, where, as a sign to Hezekiah, the Lord made the shadow retreat ten degrees. Had this been done by turning the earth ten degrees, the suddenness of it should have been apparent to the senses.[242] Thus did Amicus tacitly assume that God achieved his purpose by causing the sun to retreat.[243]

Finally, mention must be made of the passage in Joshua 10:12–14, where Joshua commanded the sun to stop in mid-heaven for nearly a day. Because it was the sun, not the earth, that was halted by Joshua's command, Amicus, Cornaeus, and Aversa saw this as powerful evidence in favor of the earth's immobility.[244] Nicole Oresme and Galileo had earlier exercised their exegetical talents on this famous passage. Aware that Joshua had commanded the sun, not the earth, to stand still, Oresme had nonetheless argued[245] that the same effect could have been achieved by causing the

---

7. His going out is from the end of heaven, And his circuit even to the end thereof: and there is no one that can hide himself from his heat." See also Amicus, ibid., 290, col. 2 and Aversa, ibid., 142, col. 1.

[238] Aversa, *Philosophia,* 231, col. 2. Riccioli also cited Job 26:7 (*Almagestum novum,* pars posterior, 480, col. 2).

[239] *Philosophia,* 232, col. 2.

[240] For this line from the Vulgate, Aversa has: "Deus fundavit orbem immobilem," where the Latin Vulgate has "ipse enim fundavit orbem immobilem." In his citation of this text, Riccioli (*Almagestum novum,* pars posterior, 480, col. 2) agrees with the Vulgate.

[241] Aversa has "Firmavit orbem terrae qui non commovebitur," which agrees with the Vulgate. Although Riccioli (*Almagestum novum,* pars posterior, 480, col. 2) starts with "Etenim," his citation of Psalms 92 also agrees with the Vulgate. In his discussion of the possibility of the earth's axial rotation, Nicole Oresme (*Le Livre du ciel et du monde,* Bk. 2, ch. 25) also found occasion to quote this line (see Grant, *Source Book in Medieval Science,* 506).

[242] Amicus, *De coelo,* 290, col. 2-291, col. 1.

[243] In Isaiah 38:8, where the same event is described, the sun's motion is made explicit in the following lines: ". . . I will bring again the shadow of the lines, by which it is now gone down in the sun dial of Achaz with the sun, ten lines backward. And the sun returned ten lines by the degrees by which it was gone down." It was this passage that Aversa cited (*Philosophia,* 142, col. 1). Riccioli (*Almagestum novum,* pars posterior, 480, col. 1) cited both passages.

[244] Amicus, *De coelo,* 290, col. 2; Cornaeus, *Curriculum philosophiae peripatetici,* 536; and Aversa, *Philosophia,* 142, col. 1. Riccioli also cited it in *Almagestum novum,* pars posterior, 480, col. 1.

[245] *Le Livre du ciel et du monde,* Bk. 2, ch. 25 in Grant, *Source Book in Medieval Science,* 507–508, 509.

earth's rotation to cease and he even suggested that the latter hypothesis was the more attractive.[246] But how could one ignore the plain intent of the text, which speaks of the sun (and moon) being stopped, but not the earth? An obvious solution was to avoid a literal interpretation, which is the advice Oresme proposed. The Joshua passage, he declared "conforms to the customary usage of popular speech just as it [i.e. the Bible] does in many other places, for instance, in those where it is written that God repented, and He became angry and became pacified, and other such expressions which are not to be taken literally."[247] In a similar manner, Galileo insisted that "to attribute motion to the sun and rest to the earth was therefore necessary lest the common people should become confused, obstinate, and contumacious in yielding assent to the principal articles that are absolutely matters of faith."[248]

With the condemnation of the earth's motion in 1616, the argument that Scripture deliberately concealed physical and other truths in order to facilitate the understanding of the common man became untenable. Scriptural passages that spoke of the earth at rest in the center of the world, or the sun moving around it, were, thereafter, to be taken literally. To say that "Scripture speaks according to the sense of the common man and not according to the truth" was, in Aversa's judgment, nothing less than "abominable," as indeed it was to most of his scholastic contemporaries who offered public opinions. Without hesitation, Aversa concluded that "for the safety of the faith, the opposite opinion"—that the earth does not rest at the center of the universe—"cannot be tolerated."[249] The many passages in favor of the traditional cosmology now took on an even more formidable aspect. No such passage could be defended by any explanation that required abandonment of the literal meaning of the text. The more relaxed liberal and allegorical interpretations of the Middle Ages were no longer tolerated. Sacred Scripture, with its many passages favorable to an immobile and central earth, became the most potent weapon in defense of the traditional geocentric cosmology. All other phenomena, whether astronomical or physical, were inconclusive. For as Koyré explained in his

---

[246] In the section cited in the preceding note, Oresme applied the same reasoning to Isaiah 38:8 explaining that although it appeared that Joshua stopped the sun and that the sun returned in the time of Hezekiah, "in fact, it was the earth which stopped moving in Joshua's time and which later in Hezekiah's time advanced or speeded up its movement; whichever occurrence we prefer to believe, the effect would be the same."

[247] I have added the bracketed words.

[248] "Letter to Madame Christina of Lorraine Grand Duchess of Tuscany," in Stillman Drake (trans.), *Discoveries and Opinions of Galileo* (Garden City, N.Y.: Doubleday & Co., 1957), 200.

[249] Aversa, *Philosophia*, 5, col. 2. Cornaeus, *Curriculum philosophiae peripatetici*, 536, makes much the same declaration. Copernicans, he explains, say that Scripture should be accommodated to our manner of speaking and feeling, so that the earth is only apparently at rest. Cornaeus insists, however, that we follow St. Augustine and always interpret the Bible literally unless "manifest reason and necessity" dictate otherwise.

Authoritative appeals were also made to the Church Fathers, though to a lesser extent. For a few such references, see Amicus, *De coelo*, 291, col. 1.

analysis of Riccioli's arguments,[250] no one "has been able to demonstrate that the earth is at rest. Indeed it is impossible to do so as in both cases—whether the Earth moved, or not—all the phenomena available to us, all the phenomena observable by us would be exactly the same."[251]

---

[250] Koyré, "A Documentary History of the Problem of Fall from Kepler to Newton," 395.

[251] Koyré should have stopped at this point, but unfortunately went on to say that "To find a difference we should look at the Earth from outside. But we cannot do it." Today, of course, astronauts and cosmonauts frequently "look at the Earth from outside" but, no more than their medieval and seventeenth-century predecessors, are they able to determine kinematically whether or not the earth really moves. The relativity of motion renders all attempts at such a determination futile.

CONCLUSION

"In science as in war," it has recently been said, "history is written by the victors. Those who first embraced a new science are styled as precursors of the latest orthodoxy. Those who stubbornly clung to the old are featured as historical curiosities. One group is absorbed, the other is absurd."[252] Although this perceptive description was formulated for Christian anti-Darwinians in their struggle against Darwinism, it applies with even greater force to the scholastic Aristotelians who opposed Copernicanism. In this momentous struggle, the vanquished paid the ultimate price: banishment from the pages of history and consignment to virtual oblivion. After the seventeenth century, scholastic cosmological treatises were little read. By the twentieth century, they were not even read or studied by historians who knew little more about them than the few arguments that had been refuted by the victors and therefore accidentally preserved.

In an age in which previously neglected aspects of history have been brought into the mainstream of historical research, the near total neglect by historians of science of scholastic arguments in defense of Aristotelian cosmology and against the rival Copernican system is indefensible. We can no more afford to ignore the losers in the struggle between the geocentric and heliocentric systems than we can afford to exclude the phlogiston theory from the history of chemistry or the Biblical concept of the fixity of species from the history of biology. To rectify a serious deficiency in our understanding of the triumph of Copernican cosmology, it is time to study the system it displaced only after a long struggle. Toward this end, I have sought in this study to describe the scholastic Aristotelian defense of the most important features of Aristotelian cosmology: the centrality and immobility of the earth. From the arguments recorded here some rather interesting, if tentative, conclusions may be drawn about scholastic attitudes toward the old and new cosmologies.

One is struck by the lack of any monolithic approach. Although scholastic Aristotelians agreed on the centrality and immobility of the earth, they presented a variety of arguments for each of these positions. While a comparison of any two sets of such arguments usually reveals some overlap and duplication, it is more customary to find divergent presentations. Nor

---

[252] James R. Moore, *The Post-Darwinian Controversies: A Study of the Protestant Struggle to Come to Terms with Darwin in Great Britain and America 1870–1900* (Cambridge: Cambridge University Press, 1979), 114.

were scholastic authors as dogmatic as frequently depicted, as is evident by the adoption of the Tychonic geoheliocentric astronomy by numerous Jesuits and by Riccioli's generous treatment of certain arguments favorable to heliocentrism. Moreover, where scientific issues arose in areas relevant to cosmology but which did not directly threaten the foundations of traditional Aristotelian cosmology, scholastic authors were not reluctant to adopt and absorb new ideas and theories, as we find with the concept of the terraqueous sphere.

But even if we confine ourselves to the arguments directly relevant to the possibility of the earth's various motions, a subject that produced severe criticism of scholastic natural philosophers, the situation is not as usually depicted. Until its repudiation near the end of the seventeenth century, the heliocentric system was contested more on physical and cosmological grounds than on its astronomical merits. Before Newton's theory of gravitation made physical sense of heliocentrism, no arguments presented in its favor were formidable enough to render traditional geocentrism untenable. Because the case for a rotating and orbiting earth had not yet developed to the point where it eroded confidence in the alternative position, scholastic arguments in favor of an immobile earth at the center of the universe continued to command widespread support through much of the seventeenth century. The arguments in defense of an immobile, central earth ranged over a wide spectrum. Many were quite traditional, resembling those that had been formulated by medieval natural philosophers. The arguments of Amicus, Aversa, Mastrius, and Bellutus fall into this pattern. Although some of them were aware of more modern arguments, their responses were often derivative and unilluminating. Other scholastics, however, were not only familiar with the traditional arguments and well informed about Copernicanism and the various proofs and experiences that had been formulated by its major defenders, but they also possessed a degree of technical competence gained from a serious study of one or more sciences. Within this group belong Riccioli, Clavius, and perhaps Cornaeus,[253] with Riccioli the preeminent figure. Not only was Riccioli familiar with the works of Brahe, Kepler, Galileo and other moderns, but he was one of the few scholastic authors who considered how the assumption of a rotating earth might affect projectile motions directed along the four cardinal directions.

We may reasonably conclude that scholastic Aristotelians of the late sixteenth and seventeenth centuries were a diverse group about whom no easy generalizations are warranted. Prior to the triumph of Newtonian science toward the end of the seventeenth century, most scholastic Ar-

---

[253] Among the more than twenty works attributed to Cornaeus in Sommervogel's *Bibliothèque de la Compagnie de Jésus* (see above, n. 34) only the *Curriculum philosophiae peripateticae* appears relevant to science. It is, however, a more technical treatise than those by Amicus, Aversa, Mastrius, and Bellutus cited in this study.

istotelians believed that the earth lay immobile at the center of the universe. In the absence of any compelling evidence to the contrary, this was not an unreasonable or untenable position. The condemnation of both the Copernican theory and its most persuasive supporter, Galileo, undoubtedly deterred some scholastics from abandoning geocentrism, but it cannot alone explain the continued support for the old cosmology. Aristotelian geocentrism was the system they knew best and with which they were most comfortable. Until the Newtonian theory of gravitation took hold in the late seventeenth century and provided a sound physical basis for the heliocentric system, most scholastics found little reason to abandon a whole complex of traditional interpretations that had served reasonably well for nearly five centuries in order to embrace what had yet to be clearly demonstrated. Scholastics may have been overly conservative with regard to the new cosmology, but they were not thereby stubborn reactionaries immune to all appeals to reason and experience. With the publication of Newton's *Mathematical Principles of Natural Philosophy* in 1687, the situation changed radically. After that famous date, scholastic authors who continued to uphold the old cosmology did so not on scientific merit, but to comply with theological decrees. Aristotelian cosmology had now lost all credibility and gradually faded away.

* * * *

I am grateful to the Program in History and Philosophy of Science of the National Science Foundation for its generous support of the research presented in this study.

INDEX